Colonize Space!
Open the Age of Reason

Proceedings of the
Krafft A. Ehricke Memorial Conference,
June 15–16, 1985

Colonize Space!
Open the Age of Reason

Proceedings of the
Krafft A. Ehricke
Memorial Conference,
June 15–16, 1985

Sponsored by
the Fusion Energy Foundation
and the Schiller Institute

New Benjamin Franklin House
New York, N.Y.

Colonize Space! Open the Age of Reason
©1985 Schiller Institute
FIRST EDITION

For more information:
Fusion Energy Foundation
P.O. Box 17149
Washington, D.C. 20041-0149
(703) 689-2490

Schiller Institute
P.O. Box 17735
Washington, D.C. 20041-0735
(703) 777-9401

ISBN 0-933488-41-6

Cover design: Virginia Baier

Front cover illustration: Detail from a painting by
Christopher Sloan of fusion reactors on the Moon, based
on the concepts of Krafft A. Ehricke. Depicted is the
latest in a row of spherical fusion power plants under
construction.

Back cover photograph by Kiyoshi Yazawa.

Printed in the United States of America

Table of Contents

Editors' Note

This book presents the proceedings of an extraordinary conference held June 15–16 in Reston, Virginia, in memory of space scientist Krafft A. Ehricke who died in December 1984. The Fusion Energy Foundation and the Schiller Institute convened the conference to bring together a group of international military, scientific, diplomatic, and community leaders who would take responsibility for solving the profound crisis gripping the Western world.

Titled "The Age of Reason in a World of Mutually Assured Survival and Space Colonization," the conference discussed scientific breakthroughs in the beam defense program and the classical scientific method that led to those breakthroughs. Proposed was a crash program approach to rescue mankind from disaster by lifting the eyes of the world to the stars—achieving a successful beam defense program as a first step along the way to industrializing and colonizing the Moon and Mars.

In the spirit of Krafft Ehricke's cultural optimism, the conference participants insisted that even today's strategic crisis can be overcome—if we create a new renaissance. As a first step, these con-

ference proceedings were rushed into print for use
by participants and others not present to begin to
get the job done.

The proceedings were transcribed from tapes.
A few papers were translated by the conference
organizers. These include the remarks of Rolf Engel, Gertrude Nebel, Hermann Oberth, and Sen.
Vincenzo Carollo.

Please note that brief descriptions of all partic-
ipants appear on pp. 381–384. Participants whose
presentations or greetings were read at the con-
ference but who were not present include Dr. Willy
Bohn, Rolf Engel, Gen. J. Bruce Medaris, Gertrude
Nebel, Prof. Hermann Oberth, Prof. Dr.-Eng. Harry
O. Ruppe, Dr. Jürgen Todenhöfer, and Kiyoshi
Yazawa.

We think Krafft Ehricke would have been pleased
with this conference—with its discussion of the
frontiers of science, from biophysics to astrophysics, with its commitment to rescue the Western alliance, and with its determination to realize the
Strategic Defense Initiative as a first step along the
way to achieving Ehricke's plan for industrializing
space.

Ehricke devoted his life to providing man with
the knowledge and tools to achieve space colonization and expand man's universe. "We must be
realistic," Ehricke wrote in 1957, "but there is a
wrong kind of realism, timid and static, which tells
man to live for his existence alone and not to rock
the boat. The kind of realism we need is the realism
of vision."

This conference made concrete the kind of vision Ehricke saw as necessary to fulfill man's "ex-

traterrestrial imperative." As Schiller Institute
president Helga Zepp-LaRouche told the 450 persons present, "I think we cannot honor him in a
better way than trying to be like him."

Susan Welsh
Marjorie Mazel Hecht

June 23, 1985

"The world of modern industrial Man is no more closed within the biosphere than it is flat. Preservation can not be limited to the environment at the expense of human growth. Human growth must aim at nothing less than the achievement of a humane living standard for all. The preservation of both environment and civilization hinges on technology and its translation into industry. Many technologies are needed to overcome the present apparent limits to growth. But the one underlying, ubiquitous technology that makes many other industrial technologies possible (either directly or by spin-off) is space technology."

—**Krafft A. Ehricke**

Krafft A. Ehricke

A New Science Mobilization Policy to Solve Today's Strategic Crisis

Morning Panel
June 15, 1985

HELGA ZEPP-LAROUCHE

Krafft Ehricke and the Age of Reason

Dear Ladies and Gentlemen, honorable guests,

I want to welcome you to this Krafft Ehricke Memorial Conference. We have assembled here in this room today people from all parts of the world, from Western Europe, from Asia, from Latin America, and the United States. We have come together in this very dangerous moment of history, when civilization is threatened in many ways, not only one. The idea of the Krafft Ehricke Memorial Conference was to bring together those people who can unite and be a force to turn history around, and create the necessary combination of forces to actually develop what is historically necessary. That is, among other things, a crash program in the style of the Peenemünde, Apollo, and Manhattan tradition for the Strategic Defense Initiative.

But we are also here to work out the concepts which are needed to recreate the conceptual basis

3

of the Western alliance on the basis of equal partnership and respect for the national sovereignty of
all countries. And hopefully our effort will create
the conditions to overcome this present danger. We
are also motivated by a fundamental belief in the
ability of men to actually accomplish the Age of
Reason. Let me say at this point that such an effort
is most appropriately in the name of, and in honor
of, Krafft Ehricke, who was not only one of the
pioneers and fundamental contributors to rocket
technologies in our age, but had been—and I had
the honor and the luck to know him personally—
one of the finest human beings I have ever met.
He was motivated by a tremendous cultural optimism, that, indeed, men could reach the Age of
Reason through space technology.

Krafft Ehricke himself was what I would call, on
the basis of Friedrich Schiller, a beautiful soul. He
was not only a member of the board of the Schiller
Institute, but he himself in his character and his
personality represented the ideal of man Schiller
was praising in his humanist ideal.

The main strategic problem that we have to discuss today and tomorrow is the fact that in the last
10 years, the period of the so-called détente, the
United States, and with it the West in general, have
nearly been strategically outmaneuvered. During
the years of détente, where the West was more or
less sleeping, the Soviets went for a steady and very
dangerous buildup. The Ogarkov Plan is a war plan,
and there is no question that the Soviet Union right
now is aiming at world hegemony, probably by the
year 1988. And even though it is very clear that
they would prefer to reach world hegemony by

political means, if they can, there is no question that they have at the same time in preparation, and are in, a full-fledged war mobilization, so that if the political way does not function, they are prepared for the military option.

Their key focus is to decouple Western Europe from the United States. Once they have accomplished this, you just have to calculate the industrial potential of Western Europe, and add that to the industrial capacity and the labor power of the Warsaw Pact, to figure out that there will be only one superpower left. We have seen in the recent elections in Greece, what the Soviets mean by grabbing entire countries. Greece right now is nothing but a puppet of the Soviet Union. It is de facto out of NATO; the only reason it is still in, is to have blackmail power, and maneuverability, within NATO.

But the key focus for the Soviet Union remains West Germany, which the Soviets, since Lenin and since the socialist revolution in 1917, have regarded as the key to conquering the world. The way they are going about it, is to say, go the way of the Socialist International, the Social Democracy, which just spent one week in East Berlin, and met with the SED (the Socialist Unity Party, ruling party of East Germany), after which they announced a historical breakthrough in security partnership. Well, if the West, especially Western Europe, submits, and says we accept your protection, the Soviets are friendly enough to say, we will not go for war. But that means Finlandization and Afghanistan. From our point of view this is not acceptable.

The media in the United States have not reported about the recent developments in the Soviet

Union. I just want to touch upon some, namely the
Gorbachov speech in front of the Central Committee, which is nothing but a complete war economy mobilization, and in addition, a Stalinist purge.
For the first time since the ousting of Beria in 1953,
Gorbachov has named ministers and party officials
as having failed in economic development. Given
the history of the Soviet Union, and the new emphasis being put on the historical role of Stalin,
people should remember what these purges were
like.

A Soviet refugee called Voslensky just wrote in
the German paper *Die Welt* that the Soviets have
put their entire war machine on launch on warning.
The West German parliamentarian Todenhöfer,
who was so friendly as to give us a message to this
conference since he was prevented from appearing
personally, just said that it is a total and complete
scandal that the Soviet Defense Minister Sokolov
gave orders to the defense ministers of all East bloc
states, to participate in a division of labor to produce
the Soviet version of "Star Wars," or beam weapons.
It is a total scandal that, despite the fact that the
Soviets have been working on their version of beam
defense technology for 10 years, the Western allies
are still debating whether there should be such an
approach or not.

The Future of Man in Space

Therefore the morning panel will discuss these issues at length. It will discuss especially the necessary
relationship between the SDI, and what we have
called the TDI, the Tactical Defense Initiative,
namely the part of the effort which would be taken

over by the European allies in order to deal with the specific threat which peoples in Europe are facing—namely, the short- and middle-range missiles which require a slightly different technology, but should be done in a division of labor with the United States. This also has tremendous importance for those countries which are not part of NATO, but are part of the developing countries, or neutral countries, which still have the desire to remain neutral, or nonaligned, and which do not want to be swallowed up by the Soviet Union.

In the afternoon we will discuss both the specific life and the contributions of Krafft Ehricke, which are very much interwoven with the history of rocket development and the history of Peenemünde. The method of science which led to the possibility of men going into space, is something worth thinking about. Nowadays with science fiction movies, "Star Wars," and so forth, people think it's quite normal for people to go into space. But you have to capture the imagination, and actually imagine what a tremendous breakthrough it was, for men to even think about leaving Earth. It is not a self-evident idea at all, that men would actually develop rockets to go into space. It is a method of thinking, and a conception of the historical role of men in the universe, to develop these ideas.

Actually this method of thinking is based on the entire scientific tradition of Western Europe. It goes back to Nicolaus of Cusa, Leonardo da Vinci, Riemann, Gauss, and the person who conceptually worked out the entire theoretical basis for 50 years of space travel to come, was none other than Hermann Oberth, who, in the 1920s worked out for the first time not only the idea of shooting rockets into space, but also the idea that men could be in such a rocket, and actually survive, and not be injured by the flight.

It is very important to study the specific scientific tradition through which this was accomplished. Oberth was helped by Busemann's work in aerodynamics, and especially the Prandtl school of the Göttingen University and Riemannian physics. The reason why we have to study this specific scientific tradition today—which is not necessarily common knowledge even among the scientific community anymore—is because only if we make a renaissance and a revival of this science, can we actually create the necessary creative outpouring that can realize the SDI in time.

In 1929 Oberth wrote a book which not only described in great detail what men's space flight would look like, but which took the same approach we have today, namely that the economic spinoff effects of space travel would be so big, that actually it would not only pay for itself, but there would actually be an economic renaissance because of this. He was such an inspiring person that he recruited many young brilliant minds around him, one of whom was Krafft Ehricke. One thing which excited these young people a lot, was a movie which was produced with the help of Hermann Oberth, *The Woman in the Moon,* from which we will show you at least some parts today. Tonight you have the possibility of watching the entire movie. The thing which is very striking, when you look at this 1929 movie about a woman and some collaborators of hers who enter a space ship and fly to the Moon, is that the designs were so absolutely similar to those used 40 years afterwards in 1969, when the Apollo 11 crew for the first time stepped physically on the Moon. You see that for 50 years, nothing much has happened conceptually beyond that. You actually grasp the pioneer spirit which these people had.

The Strategic Necessity

One can say that the period between 1923 and 1932 was the time in which the entire theoretical basis of our present space technology was laid out. Then, because of the world depression, the effort was stopped. Unfortunately it was only under conditions of war in the Peenemünde program, that these theoretical ideas were then put into effect in such a magnitude that the first rocket could be thrown into outer orbit in 1942, which was the beginning of the space age. Dornberger, who was the leader of Peenemünde, actually said at that point that these ideas of a peaceful conquest of space would never have been possible, because a centralized decision was necessary for such a large effort.

Nevertheless, I want to state this: The space scientists were not motivated by war or rockets; they wanted to accomplish what is man's duty, to conquer space. But unfortunately, there never were financial means to do that unless there was a military reason. First, it was the Second World War; then it was the Sputnik shock in 1957, when all of a sudden the United States remembered, oh, we have the scientists here, let them get to work. And today we are in a very similar situation, since we face the shock that the Soviet Union may create another Sputnik breakout in a very short period.

Therefore we are doing what the scientist fathers and grandfathers of the space technology did in 1939, when they called a day of wisdom. They called 36 professors of physics, chemistry, biology, and other scientific areas all together. They said, let's have a brainstorming session, let's put all our creative input together. Let's declare today such a day of wisdom.

The problem we have to solve is that the SDI, which in our view is the only way to prevent Western civilization from going under, is absolutely not guaranteed, and we have to fight for it. Kissinger and other people of his political conviction say, let's whittle it away, let's negotiate it away. And right now we are very, very far away from the kind of crash program which would actually be needed. The problem is that unless we do the kind of crash program which means putting to work the entire capacity of laboratories, scientists, and engineers, we will not get out of the economic crisis. The world has already plunged into a crisis which is characterized by starvation and epidemics threatening to wipe out entire continents. The world economy has plunged so deeply into a crisis that it requires new technologies, and a new technological revolution to overcome the starvation crisis which we are facing.

Krafft Ehricke correctly said that space technology, because of the complexity of the situation, is the only, and most important, fundamental technology because of the things it sets into motion either directly or indirectly to solve problems economically on our Earth. And he was also convinced, and so am I, that only through space travel, only when man lifts up his eyes away from the Earth, looks into the stars and actually thinks what his role can be, can he achieve what Schiller called the dignity of men. And only if we start to think about space, and the colonization of space, will the Age of Reason that the great humanists of European civilization were thinking of accomplishing be possible. That was the belief of Schiller, that was the belief of Krafft Ehricke: the fact that man is capable of reason, even under the most horrible condition of crisis. This is our most fundamental belief.

Man is capable of reason and overcoming every crisis. And in this spirit I want to open the Krafft Ehricke conference, and welcome you.

Greetings to the Conference

Rolf Engel

I met Krafft Ehricke for the first time in late Autumn 1942 during the discussion with Dr. Walter Thiel in Peenemünde. Although he had been called away from the front only a short time before, he had quite evidently worked himself into the complicated material of fluid rocket motors very quickly. After that meeting, we lost sight of each other for a time. Even after the war, I only knew of Krafft Ehricke that he was working in the United States. It was at the annual meeting of the International Astronautics Federation that I heard of him again. Very soon, Krafft Ehricke became one of the most interesting speakers at our conferences. He delivered excellent reports, always developed with considerable didactic talent, and always looking far into the future. His presentation, however, never went beyond real technological potentials, although the subjects he spoke about sounded like titles of science fiction books. He remained irreproachable scientifically, because he was a master of rocket technology and space flight. His departure from the ranks of space activists is a heavy loss for all of us. Krafft Ehricke we shall never forget.

Gertrude Nebel

Allow me to convey my heartfelt greetings to your conference.

Throughout his life, my husband worked to realize the age-old dream of mankind: to fly to the Moon, to venture out into the universe. The development of rocket technology, which laid the crucial foundations for space travel, was for him first and foremost a contribution to universal scientific development, and thereby to the preservation of world peace.

For him, it was self-evident that underdevelopment, hunger, and sickness, which now have once again assumed a threatening worldwide scope, could be overcome once and for all, if the achievements of space travel could be used to benefit all men and all peoples.

In the hope that this conference will contribute to furthering these conceptions of my husband, I send my friendly greetings.

Dr. Hermann Oberth

I have a great interest in the conceptions of my very dear and esteemed colleague Krafft Ehricke for the exploitation and settlement of the Moon and space by men.

I have myself devoted considerable thought, earlier, to the exploitation of the Moon for industrial and astronautical purposes. The Sun ought to be used during the 354 Moon-day phase for continuous operation of electrical-power installations.

Shots into space should be effected with electro-magnetic accelerators, or slings, and the energies required for the long Moon-nights and the peak-use periods of accelerating payloads, should be charged as follows: A series of strong-wall metal containers, filled at the bottom with fluid, and at the top with compressible gas (materials for this purpose depend upon those found on site on the Moon), are to press the fluid through appropriate turbines, into similarly formed empty containers.

My conception of the Moon accelerator, or Moon sling, is that of a sufficiently long horizontal accel-erator-rail, which drives a magnetic car, which is open at the front or can be easily opened. Braking energy will be produced, as far as possible, by dynamos, and used to drive the fluid back into the containers which have become empty. The velocity by which the freight leaves the magnetic car ought to be approximately 2,320 m/sec. If the velocity is 2,540 m/sec., the freight would fall to Earth.

The distance required for braking, decelerating the freight, ought to be as short as possible. The end of the sling ought to be so positioned, that the freight stops at the far-end 60 degree libration point of the Earth/Moon system, where the freight is caught and transported to its final destination. Structures in space can then be assembled out of their components at the appropriate places in space. Krafft Ehricke's idea is very good, to decelerate vehicles that are to land on the Moon on a skid-track of Moon-sand. I had never thought of this method of braking the freight velocity. Hopefully it will work.

The more I think about the settlement of the

Moon and the future of mankind in space, the more I discover, that it is improbable that we still have the time to undertake such things. Mankind would have to learn to re-think philosophically and psychologically. It is paradoxical, for example, that the expenditure of $25 billion for the Moon-landing is considered a waste of money, because people still go hungry, but no one makes the same criticism of the $600 billion expenditures for arms each year.

I am of the same opinion as my colleague, Krafft Ehricke, that space flight technologies can overcome all of the apparent limits of our present existence on Earth—lack of energy, scarcity of resources, environmental pollution, scarcity of food—and provide for an expansion of human population. This would presuppose, of course, that the willingness to conduct war in our democratic era immediately ceases, and it becomes possible to create the conditions of justice, which alone are guarantees of peace. To stop the willingness to conduct war, knowledge and the correction of psychological and sociological aberrations are necessary.

Were I to report from my own experience, why the first and second world wars happened, the chief cause that crystallizes is the *political lie*. Claims were made which did not correspond to truth, but remained uncontested, or that these claims had been refuted was unknown to humanity in large part. I am, after all, 91 years old, and have often held positions, which gave me deep insights into political events. The psychological preparation for war is once again to be seen everywhere. The two power blocs in the East and West mistrust each other fundamentally, for the media-bosses only let that news through which proves the degeneration and evil of the adversary. The only common bond is that of hate they both bear against their former adversary Germany. This hate is cultivated by both sides. Recent events provide impressive material for study in this regard.

Although they were encouraged by some reasonable foreign commentators, official representatives of the German people did not dare to insist upon highlighting historical events in their true relationships, at least to take the burden from the shoulders of postwar generations.

If the human being is continuously harassed, his intelligence alone cannot yield an objective view of his situation. I learned this, when I visited the well-known Swiss scientist, Professor Bluntschli. I wanted to take the side of a former student in an argument, and we set a discussion date. When I arrived, he greeted me with a flood of lying accusations that had been spread during the world war, and which he had collected—and he told me he would never even consider doing anything on behalf of my student, and that he had invited me to see him to have the opportunity to unload everything he wanted to say upon a German. I knew that most of what he said was wrong, but when I told him that, he screamed: "Surely, you do not want to insinuate that Swiss journalists lie." "Anyone who says such things," I told him, "is wrong—even if he is a Swiss journalist."

Finally, I asked him. "Let's assume a researcher on Africa comes to a tribe, and the chief of the tribe tells him, the Swiss live in caves, eat human flesh, and clothe themselves with animal skins." As the researcher wants to refute him, the chief says: "Our medicine men here said so, and Mr. Researcher surely will not insinuate that they lie." At this Professor Bluntschli escorted me out the door.

The case of Lloyd George demonstrates that this is not an exceptional example. He wanted to create a just world at the end of the First World War. But it turned out he could not reach his goal, because incitement was stronger than the commands of reason.

I think such false insinuations and claims are especially dangerous, when German statesmen have nothing to say against them. Since, aside from the Germans, practically any people has a sense of nation, an unprejudiced person has to come to the conclusion, "there must be something to it." Those who do not combat lies are more dangerous to world peace than the enemies of Germany, of whom one knows at least that they are such.

This German attitude will lead people in the East and West to let the nuclear bombs fall on Germany, because "it doesn't matter much if this bunch dies."

That, however, the third world war, and the probable end of humanity will be the result, is unfortunately not acknowledged by public opinion makers.

I conclude:

The indispensable prerequisite for the further cultural development of mankind is the yearning for historical truth.

In war, both sides commit injustices, but it needs to be said finally that one cannot make one side solely guilty.

I only hope that we shall have enough time to

raise the sails to realize Ehricke's vision of "Homo Sapiens Extraterrestris," so that man may leave his berth in the flaming harbors of the Earth, and steer a new course into the world of unlimited growth.

Prof. Dr. Harry O. Ruppe

We have lost a great scientist, a great humanitarian philosopher, and a great human being. Krafft worked for more than 40 years to create, shape, and direct astronautics.

From combat service in World War II, he—a graduate aeronautical engineer—joined the legendary Peenemünde team in '42; with that group, he came to the U.S.A. in '47. In '52 he left for the Bell Aircraft Corporation, where Dr. Dornberger was working. Two years later he teamed up with Convair's Atlas design team. Finally in about 1979, he founded his own consulting company "Space Global." On Dec. 11, 1984, he died after prolonged illness.

His professional accomplishments are many; I can touch on a few highlights only:

- The first industrial study contract for nuclear rocket propulsion, 1942;
- The concept of a space station essentially built from last stage tankage (finally, Atlas vehicle) 1954-1958;
- From the same time, a winged Earth landing vehicle;
- In about 1956, the extremely low-mass solar-light-concentrating "Ehricke Sphere" design;
- His series of books, *Space Flight* (v. Nostrand, 1960 and 1962; Vol. III, unfortunately, never appeared in print), are highly valuable even today,

both in factual information and in ideas touched upon. I would only wish those books were studied more widely;

● The Centaur high-energy upper stage was strongly influenced by him (1963); more than 20 years later, it is still the best vehicle of this class, with non-diminishing program importance;

● Krafft was a prolific writer, both in all learned astronautical journals and in the public press. He was an outstanding idea-generator throughout his life. I was privileged to work with him during the first half of the sixties on the project of manned interplanetary missions—and no new ideas have emerged since then!

More than the last decade of his life was devoted to studies of settlement of near-Earth space, including our Moon, for the peaceful benefit of mankind. His "Space Imperative" and "Open Earth" concepts are countering the gloomy pessimistic expectations of the Club of Rome.

E. Sänger pointed out that sometimes membership of an individual honors the academy—no doubt this holds for Krafft, when in 1960 he was elected Member Section 2 of the IAA [Institute of Astronautics and Aeronautics]. Rightfully, he was the recipient of many honors distinguishing the recipient as much as the donor.

Krafft's working power and enthusiasm, his deep knowledge, openness and kindness are a permanent challenge for all of us. His ideas will be with us as long as "space cadets" accept the challenge of the next frontier.

Best wishes for your conference.

Kiyoshi Yazawa

Dr. Krafft Ehricke always strongly impressed me with his incomparable and passionate devotion to progress and with his great achievements in space science. I am sorry that I cannot be at this important conference in his memory.

I believe that it is my role, as a journalist and editor, to convey on every occasion to the Japanese people what Krafft Ehricke has done and what he wanted to be realized for the heart-lifting future of mankind.

DR. JUERGEN
TODENHOEFER

SDI—With the Europeans

1) The Chancellor of the Federal Republic of Germany, Dr. Helmut Kohl, has the full backing of the Christian Democratic caucus of the German Bundestag for his fundamental affirmation of the goals of the American Strategic Defense Initiative, and his declared policy that the Federal Republic of Germany should participate in SDI research.

2) SDI is a research program for laser- and particle-beam weapons, which is to demonstrate to what extent the utilization of these new technologies can alter the previous "offensive deterrence" into a "defensive deterrence." If the research succeeds in making such a breakthrough, it would have far-reaching effects upon the strategy of the Alliance. The question is thus posed, whether the European NATO-partners, and particularly we Germans, can stand aside from these changes?

3) The Soviet Union is the only country in the world that possesses a ground-based, nuclear mis-

sile defense system against ICBMs, located in the area around Moscow. Moreover, the Soviet Union is improving this defense technology, and expanding its production capacities for new developments in these technologies rapidly. There is evidence that Moscow intends to defend its most important military installations with ground-based missile defense systems within a few years. One part of these systems will be mobile-stationed, and thus difficult to discover by means of Western reconnaissance satellites. Furthermore, the Soviet Union has been working intensively for a number of years on a space-based missile defense system, based on laser and particle-beam weapons. The Soviet Union has expended several times more financing in this area than the corresponding American program, and leads the United States in its research results.

4) Parallel to these efforts to construct a missile defense, the Soviet Union is engaged in a massive program of modernizing its offensive weapons. The Soviet Union currently produces 350 nuclear ICBMs per annum, 150 of which are ground-based, while the United States has had considerable difficulty in obtaining at least 100 new MX-ICBMs. This discrepancy in both the offensive and the defensive areas makes an American reply imperative.

5) The prerequisite to a decision in favor of constructing a space-based defense system of the United States against ballistic nuclear missiles must, of course, be that the conclusion of the initial five-year phase of SDI research demonstrates that:

a) SDI is technologically feasible with reasonable financial allocations;

b) Europe must be just as securely protected from ballistic nuclear missiles as the United States; and

c) Military stability between West and East must be increased, and not reduced, by SDI.

If even one of these conditions is not fulfilled, the alliance partners of the United States will hardly be able to agree to American SDI plans. But it is hardly to be expected, in any case, that the United States would continue to pursue realization of SDI were these conditions not met.

6) Objections raised to West European participation in SDI, based on the concern that we might receive too many contracts and too much work, are totally new to me. I have confidence in the energies and ingenuity of German firms.

7) We have sympathies for the idea of the French President, Eureka; but Eureka is not an alternative to SDI. Eureka cannot replace participation in SDI. We will therefore do our utmost to achieve a participation of the most important industrial nations of Europe in the SDI research program. We will seek the broadest unified standpoint of Europe possible, in order to give SDI-research the broadest possible European support.

8) The West, however, should entertain no illusions. The Soviet Union will initiate the grandest propaganda campaign of the postwar period, in the attempt to destroy the American SDI plans, without giving up or restraining their own missile defense plans. The campaigns against the neutron bomb and the Pershing II/cruise missile modernization of nuclear forces in Europe were but a harm-

less foretaste of the Soviet propaganda campaign we will now face. The only means to use against this propaganda offensive is a broad-scale offensive information policy on the part of the West. One who has a strategy for defense, as the West does, must wage an offensive information policy.

9) The West must also refrain from belittling the successes which SDI has already achieved. However one might wish to judge the SDI, and the transition from a strategy of "offensive deterrence" to one of "defensive deterrence" in the context of Flexible Response, it has already achieved one thing: in addition to the unanimity and steadfastness of the West, and especially the Federal Republic of Germany, in the issue of the modernization of NATO's nuclear forces in Europe, the SDI has brought the Soviets back to the negotiating table at Geneva. I am convinced that this will not remain the last of the successes of President Reagan's Strategic Defense Initiative.

10) If it is possible, by means of a purely defensive defense screen in space, to reliably destroy missiles flying through space, we will have achieved two of the central aims of our policy for security and peace. First, neither the West nor the East would then have the capability of conducting so-called first strikes and incapacitating strikes. Second, nuclear missiles would then only be worth their value of scrap metal, from a strategic standpoint. This would be, de facto, the grandest disarmament since the Second World War.

LYNDON H. LAROUCHE, JR.

Ehricke's Contribution to Global and Interplanetary Civilization

As each of us is born, each of us must die. Within that brief interval of life, what distinguishes a life as human, as exalted above the condition of mere beasts, is that which the individual contributes to the enduring benefit of future generations. Our beloved and most accomplished friend, Krafft Ehricke, has bequeathed to future generations a beautiful and most valuable gift.

For the information of those who have not been told, let this be said here, so that it may be repeated elsewhere. Krafft's adult life was dedicated to what became an important part of the work of a small group of dedicated pioneers associated with Dr. Hermann Oberth. These men and women, assembled amid the horrible conditions of material and moral decay following Germany's defeat in the First

World War, dedicated themselves to uplifting the moral condition of all humanity, to turn mankind's eyes from petty squabbling in the mud of this planet, to exploration and colonization of space. To this purpose, these pioneers of science, drew upon one of the most precious contributions which German culture had already given to all mankind, the scientific heritage of Nicolaus of Cusa, Johannes Kepler, Gottfried Leibniz, and Karl Gauss.

With the aid of that scientific heritage, these pioneers have enabled mankind actually to explore, and now soon to colonize nearby space. What they have accomplished, could not have been accomplished, without the advantage of the heritage of Leibniz and Gauss. They have led all mankind along the only pathway by which we might reach the stars.

In that effort, our dear Krafft Ehricke served with notable distinction, to the degree that his name must be remembered most prominently by those who construct the first colonies on the Moon and Mars. He has helped in an important and practical degree, to make clear to humanity, that it has been the intent of the Creator that mankind's destiny is to become mankind in the universe. There, in the stars, lies mankind's entry into the long-awaited Age of Reason, when our species sheds at last the cultural residue of the beast.

To the pioneers assembled around Hermann Oberth, as to all leading currents of scientific progress in modern history, came the awful truth, that no government thus far, has been able to muster itself to support generalized scientific and technological progress, except in connection with military ventures. This has been the twentieth-century history of the United States and Western Europe. It has been the history of Germany in particular.

The circles of the great Friedrich Schiller represented the highest degree of progress of understanding of the direct connection between scientific progress and the principle of pure beauty. Yet, all their efforts were frustrated, until the battle of Jena so humiliated the Prussian state, that that state reluctantly turned to the circles of Schiller, to prepare the Liberation Wars out of which every later institutional progress of Germany emerged. In the feudalistic reaction which seized Europe at the 1815 Congress of Vienna, the efforts of the Humboldts to make Germany a center of the world's scientific progress, would have been crushed, if the Prussian military had not intervened to subsidize the efforts of Alexander von Humboldt and *Crelle's Journal*.

In the United States, when the economic and scientific policies of the Founding Fathers had been all but crushed out of institutions of government, it was the war of 1861-1865 which transformed the United States into a great agroindustrial power. It was Britain's mobilizing the United States for the First World War, which produced the industrial progress of the decade following 1907. It was the mobilization for the Second World War, which unleashed the United States' agricultural, industrial, and scientific recovery from the Great Depression. It was the aerospace mobilization of the United States, up into 1966, which continued the agricultural, industrial, and scientific progress of the United States after that war.

So, it is the lawful irony of the modern history of science, that the noble passion of the Oberth group is known to the world today in terms of their military accomplishments. They are known to common opinion, not as the conquerors of space, but in terms of the military contributions of Peenemünde. They are known as the group of scientists and engineers who gave the world military rockets, the principles of supersonic aircraft, shaped explosive charges, and numerous other such artifacts. If the Soviet Union had not gotten about 6,000 Peenemünde veterans drunk, and hauled them into Soviet workshops, Moscow would not have acquired that German science upon which its acquisition of modern military rockets and thermonuclear detonations depended. Without "Operation Paperclip," the United States, too, would have had great difficulty in mastering these technologies.

It is therefore not accidental, that a unit of Soviet intelligence established by the late Suslov, has successfully penetrated a corrupted channel of the U.S. government, to convey forged Soviet libels against U.S. veterans of the Peenemünde project. The Soviet government knows very well, through its own scientific debt to Peenemünde, that the United States will be defenseless against the Soviets' massive military mobilization for 1988 now being conducted, unless the United States turns once again to the resources of aerospace development built up here around our Peenemünde veterans. So, Soviet intelligence, working through the Moscow Procurator and the East Germany-based VVN, has conduited forged documents, through known Soviet agents, into the Office of Special Investigations, for the purpose of scapegoating and demoralizing the persons who are either veterans of Peenemünde or closely associated with them.

This action by certain officials and other citizens of the United States, is very simply, very plainly, pure and simple treason, pure and simple aid and comfort to a Soviet government which has declared its mobilization for impending "Holy War" against the United States and its allies. According to documented Soviet military doctrine, the Soviet Union is already in a state of war against the United States, and under those conditions, those persons who are aiding and abetting this Soviet-directed scapegoating of Peenemünde veterans, are guilty of treason as the U.S. Constitution defines treason, as giving aid and comfort to the enemies of the United States in time of war.

My friends, we are again in a condition of warfare. Except for escalating sabotage and assassinations, in Europe and the United States, being directed by Soviet intelligence, it is not yet a shooting-war. However, Soviet doctrine specifies, that the state of war begins with a prewar mobilization up to the level of full-scale war-economy. Not only are Soviet forces mobilizing just so; that new Stalin in Gucci shoes, Gorbachov, and other principal Soviet officials, have affirmed, repeatedly, that this is the present practice and intent of that government. True, there are many wishful dreamers around Washington, who deny the simplest facts known to every European leader on this and related questions. Avoidance of the facts, does not alter the facts. We are again at the threshold of general warfare, and sane men and women will act accordingly.

So, once again, as the veterans of Peenemünde

have twice experienced this hard reality, those of us who would prefer to colonize the Moon and Mars, are condemned to devote our competencies to perfecting the instruments of warfare.

Krafft Ehricke understood this bitter truth very well. By "very well," I mean, that as my wife and I were deeply privileged to know him and to collaborate with his efforts, Krafft as we knew him was both a world-citizen and a patriot, in precisely the sense Schiller defined this quality of the beautiful soul. We must hate war, as General Douglas MacArthur hated war, but we will not buy peace at the price of the degradation of all civilization; we will not buy a peace at the price of transforming our children and grandchildren into slaves or degraded beasts.

These foregoing observations are essential, to situate both the subject matter of my report to you today, and to situate that subject matter in the common spirit we here assembled share, in reflection on the memory of our dear friend.

It is now nearly three and a half years, since I announced the design of a new strategic policy for the United States and its allies, a policy later announced by President Ronald Reagan in his famous televised address of March 23, 1983. What I outlined, and that for which my associates and I have campaigned throughout most of the world ever since, was a combined strategic and tactical defense, based upon the orders of magnitude of superiority in firepower and mobility of coherently directed electrohydrodynamic impulses. I proposed that this be accomplished by means of the kind of "crash program" we experienced most recently in the 1958-

1966 phase of the Apollo project and related aerospace developments.

Although the relevant industry of France, Germany, and Italy, among other nations, is already committed to these lines of research and development, the new policy of defense is not yet a "crash program." Once Washington belatedly awakens to the reality of the present strategic situation, SDI and its tactical complements, will be transformed into a "crash program." It is in this setting, that we today have urgent lessons to be learned from the Peenemünde experience.

Although many of the valuable lessons of the Manhattan Project and of the Apollo Project, are embedded in the knowledge of some of our military specialists and scientists today, the essence of the principles of a successful "crash program" is not competently understood. To implement the SDI and related projects through a "crash program," not only must we eradicate the obstacle of "systems analysis" from the Department of Defense. The best specialists in our military need the best insights which can be contributed into the deeper principles of a successful "crash program."

There are either assembled, or represented here, today, a body of knowledge and experience, which, in total, is best equipped to assist in providing the urgently needed answers to such questions. The connection between the primary dedication of the Oberth group, to reaching the Moon, and the military work of the Peenemünde veterans, is perhaps the best case in point from recent experience. To that point, I shall now summarize a draft proposal, indicating my best estimate of what the essential principles of the needed "crash program" are.

I stress draft proposal, to emphasize that this is subject to modification through aid of the experience represented here today. I would hope thereby, to stimulate such a discussion-process. However, although the proposal is conditional in detailed features, it has the advantage and authority of resting primarily on known principles of the current of economic science established by Leibniz, and upon the standpoints in method, successively, of Carnot and Monge in France, and of the circles of Gauss in Germany.

I summarize first, the essential historical background, and then summarize the draft policy itself.

The History of 'Crash Programs'

By "crash program," I mean the tight integration of the most advanced, most fundamental scientific research with the production and deployment of new technologies in a general way, such that there is no organizational separation between the most fundamental scientific research and production in general.

The history of "crash programs" begins with the collaborations on this matter, between Cosimo de Medici in Italy, during the rise of the Golden Renaissance. The first implementation of a true "crash program," was that led from Milan, Italy, by the collaboration between Luca Pacioli and Leonardo da Vinci. The next true "crash program," was launched by the French Minister Jean-Baptiste Colbert, following the 1653 defeat of the Habsburgs by Cardinal Mazarin. Colbert's sponsorship of Huygens and Leibniz in Paris, was the driver out of which European scientific progress was revived, and out of which the industrial revolution was directly planned and set into motion. The next true "crash program," was that attempted in France, beginning 1793, under the leadership of Lazare Carnot and the 1794-1814 Ecole Polytechnique. The development of Germany's world supremacy in science, around the central figure of Gauss, was the result of the attempt, led by Alexander von Humboldt, with collaboration of the exiled Carnot, to transfer the work of the Ecole Polytechnique, then being suppressed in France, into a safe haven in Germany. The economic reforms introduced by Friedrich List, intersecting the work of Humboldt's collaborators at Berlin and Göttingen, both backed by heirs of Scharnhorst in the German military, is the secret of the scientifically driven industrialization of nineteenth-century Germany.

The superiority of the work of the Peenemünde veterans' work in implementing crash-program efforts, might seem to be explained by the fact that the Göttingen tradition, including the Betti-Beltrami offshoot of Göttingen in Italy, is peculiarly superior to the Cartesian tradition of France and English-speaking countries in hydrodynamics. So, it should be noted, during the 1920s and early 1930s, Italy was the world's leader in air-frame design, and already a leader in the scientific principles of supersonic aircraft-design. Professor Busemann has emphasized the debt which the Oberth group had to its Italian collaborators during the 1920s and early 1930s. So, it might be assumed that Peenemünde had only a specialized competence, well-suited to rapid advances in aircraft and rocketry.

Such notions of limited competence must be cast aside, as we examine the point, that Bernhard Riemann's 1859 paper "On The Propagation of Plane Air Waves of Finite Magnitude," pertains not only to the transition to supersonic velocities, and shaped charges, but is key to isentropic compression of thermonuclear plasmas, and was also the starting point for Schrödinger's exploration of the hydrodynamical structure of the electron.

The successive heritage of Nicolaus of Cusa, of Kepler, of Leibniz, of Monge, and of Gauss, is an elaboration of the principle that physical space-time is essentially hydrodynamic in character, and that the mathematics of physical space-time must be derived by aid of a rigorous development of, and training in what is called synthetic, or constructive geometry. In the rise of German science, out of Schiller's attack on Immanuel Kant on the issue of aesthetics, the efforts of the anti-Kantian student of Schiller, Herbart, to base education upon a fusion of the classics with education in geometry, Gauss's elaboration of the implications of the arithmetic-geometric mean, and Riemann's basing his program for advancement of Gaussian physics entirely upon a correction of Herbart's error, are crucial. The explicitly anti-Kantian physics of Gauss, is based on elaboration of the principle of scientific method discovered by Nicolaus of Cusa, called today the isoperimetric principle, that only circular action, not straight-line motion of point-masses, is self-evidently existent in physical space-time.

Hence the physical space-time of Riemann, which

is the proper physics for economic science, defines the invariant characteristics of the laws of our universe as congruent with a harmonically ordered hyperspherical function, in which conic self-similar spiral action is the form of physical least action. The general method of experimentation, which flows from the Gaussian manifold, requires a synthetic-geometric construction of the indicated relations of a phase space, and the extension of this construction hydrodynamically according to principles of self-similarity.

If this is taken into account, economic science shows us how a properly defined "crash program" must work, and why the Gaussian tradition, as mediated to a large degree by the work of Prandtl, is the best vantage point for such programs.

The possibility of correlating fundamental scientific progress directly with increases of the productive powers of labor, was opened up by Leibniz's founding of economic science, with emphasis on Leibniz's defining the meaning of the term "technology," in the context of study of principles of heat-powered machines. Instead of accepting a Cartesian scheme, in which straight-line motion of point-masses is axiomatic, economic science consistent with the principle of least action, is based upon the fact that action in physical space-time is intrinsically circular action, as Leibniz shows in his famous refutation of Descartes's errors on the notions of momentum and work.

The notion of "technology" arises in elementary study of the principles of heat-powered processes, by considering the simplest ideal case. In the hypothetical case, that two machines employed to produce the same quality of product, consume coal-equivalent at the same rate, consider the special case, that the operative employing one of these two machines produces greater output than employing the other. This ideal case, forces to our attention, the notion of the internal organization of the productive process as a cause of increase of the productive powers of labor. The notion of ranking such internal organization of processes, according to correlation with increase or decrease of relative productive powers of labor, is the simplest outline of a generalized notion of "technology."

Since I have elaborated this conception in several published locations, I shall not repeat those details here. I shall merely summarize those features of economic science, which bear directly on the proposition chiefly under consideration in this report as a whole.

The correlation between advances in technology and increase of the productive powers of labor, is measured by functions of the interrelations among these four categorical elements:

1) the usable energy throughput per capita and per square kilometer;

2) the energy-flux density of the power supplied;

3) the capital-intensity of production;

4) the internal organization of the productive process as such.

These are the four, interdependent factors employed for measuring accurately the relative level of technological development of compared economies. Provided that productivity is measured in units of increase of potential relative population-density, existing statistics from national and supranational agencies, provide provably accurate qualities of measurement. This measurement has the specific and more or less indispensable usefulness, of enabling us to estimate the investment-budgets needed to increase the productive powers of labor of any economy by some projected amount.

The measurement of the causal relationship between quantified advances in technology and resulting increases in the productive powers of labor, requires specific choices of method and procedures in mathematics. The method must be based on a rigorous application of the principle of synthetic geometry, and the physical space-time of economic action must be geometrically the physical space-time elaborated by the work of Gauss, Dirichlet, Weierstrass, Riemann, and Cantor.

To unify mathematically, the four interdependent aspects of technology, we must define "energy" in terms of self-similar circular action in a Gaussian manifold. We must think of energy in terms of both radian measure of perimetric action in space-time, and in terms of the areas and volumes subtended by either cylindrical or conical self-similar spiral action. Only on those conditions, can the four interdependent aspects be integrated. So, for example, we think of the measurement of energy as action, by a standard wavelength of coherent electromagnetic photon, such as a standard wavelength of perfectly lased yellow light.

There is one additional point concerning economic science, which must be stressed here and now, if the nature of well-designed "crash programs" is to be exposed. To make the point clear to those who are not professionals, I illustrate the problem which requires economic science to employ the mathematical physics of Riemann.

If we measure the relative productivity of the U.S. economy, during various intervals of the postwar period to date, we have the following picture of the relationship between technology and productivity. We make these measurements in terms of changes in potential population-density, and measure productivity in terms of per capita physical outputs.

The U.S. economy recovered from the postwar recession, with the launching of military mobilization during 1939, and stumbled along on the basis of this up-tick, until the deep recession of 1957-1958. The U.S. economy recovered rather vigorously under the combined impact of the aerospace and Kennedy investment-tax-credit impulses, into the middle of the 1960s. With the partial demobilization of research, under the "Great Society" program, the economy stagnated into 1971, and has been in continual, accelerating collapse ever since.

Look more closely at what has occurred since President Carter and Federal Reserve Chairman Paul Volcker introduced what Volcker described as "controlled disintegration of the economy," beginning October 1979.

By February of 1980, Volcker's measures sent the economy into a rather steep decline, until the summer of that year. A slower rate of decline, which some called a partial recovery, followed. During the spring of 1981, the economy went into another steep decline, into October 1982. Beginning the first quarter of 1983, the rate of decline slowed significantly, and then began to accelerate again during the spring of 1984. Since March of this year, the rate of decline has been accelerating rapidly, erupting now in the forms of a declining price of the U.S. dollar, and waves of bankruptcies throughout the banking system, as well as in agriculture and industry. This is the steepest decline in the economy since the 1931-1933 period.

Most of you have either ridden on a roller-coaster, or have at least watched the procedure. You chug up to the high point of the structure, and then begin an accelerated descent. You go up and down. Each time you go up, you reach a high point which is lower than the preceding highest elevation of the ride. Finally, you reach the bottom.

That is the way economies usually collapse. Since October 1979, the U.S. economy has been on a roller-coaster ride downhill. The brief periods which some have called "economic recoveries," during this period, were not recoveries. That is, the rate of per capita output of physical goods never reached the level of a previous high.

In the history of modern economies, general advances and declines in productivity always occur in jumps. General falls in levels of productivity always resemble a roller-coaster ride downhill, whereas general rises occur in the reverse pattern. Why is this so? The answer is elementary. At least, the answer is elementary to an economic science based on the Gaussian manifold.

To make the explanation as brief as possible, I

show you some diagrams from my article refuting
the notion of "artificial intelligence."[1]

How do we describe an economic process, in
which growth of productivity is caused by consistent
technological progress under conditions of rising
energy-intensity and capital-intensity? In first approximation of the ideal classroom case, the function we require is generated as the compounding
of conic self-similar spiral action with conic self-
similar spiral action. The result, as you see, is a
hyperboloid (Figures 1-3). This seems to present
us with a nasty problem. It seems that the arms of
the hyperbola are shooting off into Cartesian in-
finity. This is a mathematical discontinuity. Yet, we
know that economies do not come to a halt because
of successful increases in productivity.

First, we must eliminate the Cartesian absurdity.
We do this by projecting the image onto a Rie-
mannian sphere. No more silly Cartesian infinities.
I discuss this in some detail in the published item
I have indicated, and in additional detail in a pub-
lished criticism of the incompetencies in Professor
Wassily Leontief's featured report in the June 1985
Scientific American.[2] Since those items are available
for those who wish to follow up the details, I need
only summarize the most essential point now.

At each discontinuity, there is the addition of at
least one singularity to the economic phase space.

1. "The continuing hoax of 'artificial intelligence': The
multi-billion dollar boondoggle," *Executive Intelligence Review*, May 14, 1985, pp. 24-33.

2. "Wassily Leontief acts to block the effective implementation of the SDI," *Executive Intelligence Review*, June 10, 1985,
pp. 6-12.

This corresponds thermodynamically, to an increase of the energy-of-the-system of the economic process, when energy of the system is measured as potential population-density per capita. We measure action as the perimetric area of action swept by our figure, and the work done as the volume of the sphere subtended by that area of displacement. So, increasing the energy of the system per capita, means that at the point of apparent discontinuity, the continuing action occurs on the surface of a larger concentric sphere. What you see in the figure now, are projections of the action on larger spheres projected on the one sphere.

In the ideal classroom case, the concentric spheres are in a self-similar harmonic ordering, as we would expect by adding a third degree of conic self-similar spiral action to the mathematical model.

This classroom model I have outlined, conforms more or less exactly to what has occurred historically, both in periods of technological progress, and in periods of economic devolution, such as the post-1966 period of the U.S. economy's descent into what is sometimes called a "postindustrial society." The roller-coaster character of economic collapse, means, that as a period of steeper collapse destroys elements of the physical economy's farms, industries, and basic economic infrastructure, the economy drops to a lower sphere in our classroom example. It appears to stabilize itself briefly in that lower state of the system, and then collapses to a still lower sphere. In the rise of productivity, in a technology-intensive mode, the pattern indicated in the model occurs.

This model illustrates what we ought to mean,

when we observe that economic processes can not be analyzed by methods of systems of linear inequalities, because economic processes are intrinsically, everywhere nonlinear. Economic advances occur in approximately the same way Riemann describes the generation of transsonic shock waves, in his "On The Propagation of Plane Air Waves of Finite Magnitude." Let us term the singularities of economic processes, "technology waves." The radiation of the impact of newly introduced, advanced technologies, spreads through the economy in waves, altering the division of labor and productivities throughout the economy, and adding new kinds of materials and instruments to the repertoire of production as a whole.

In devolution, the reverse occurs, as we note that the United States today has lost many of those branches of industry which were essential to the Apollo Moon landing. We also observe, that the raw energy throughput and energy-flux densities of production, have collapsed per capita for the U.S. population as a whole. We are operating in a lower state of structure, and at lower per capita energy of the system, than we did in 1970, when the general and generally accelerating rate of decline began.

Now, we must reverse that decline. I review briefly the key parameters of the needed technology-driven recovery, and then come to my summary proposals on "crash program."

The New Industrial Revolution

It is a remarkable experience today, to view the Fritz Lang film of 1929, *The Woman in the Moon.*

The elements contributed to the design of that film by Dr. Oberth, show us how little we have progressed beyond the conceptions of spaceflight which German science had outlined more than 55 years ago. Especially under the conditions of relative scientific stagnation of the past twenty years, we know very well what present frontiers of science will shape the technological revolutions of the comming twenty to thirty years.

In projecting either the fulfillment of the SDI, or technological revolutions in the economy generally, we divide the required technologies into two classes, which we may term quite usefully as primary and auxiliary technologies. The primary technologies available to us for SDI and technological progress today are three:

1) control of thermonuclear plasmas as a source of very high quantities and energy flux densities of organized electromagnetic action;

2) coherent organization of directed-energy impulses;

3) what we call, in shorthand, optical biophysics: the electrohydrodynamic characteristics of living processes.

The auxiliary technologies include such improvements in computer technology as true parallel processing, and true analog-digital hybrids designed for efficient processing of the class of nonlinear functions implicit in a Gauss-Riemann electrohydrodynamic manifold. In terms of SDI and related classes of military assignments, the first two categories of new technologies are the source of the firepower and mobility of the weaponry, and the auxiliaries are needed for acquiring and aiming at targets, as well as delivering the systems to their firing positions.

To grasp the general implications of the new technologies for both the economy and military science, the most efficient view is developed by giving our "crash program" teams the mission assignment of establishing and maintaining colonies on both the Moon and Mars. In other words, if we wish to develop the SDI and its offshoots in the best way, the way to organize the program is as a by-product of a mission assignment for colonizing first the Moon and then Mars. Every technology we require for military purposes, will appear as a by-product of the primary mission assignment. The Soviets already understand and are applying that principle, which U.S. policy has so far failed to grasp.

Fusion provides us the needed technology for powered interplanetary flight, superseding the problems of unpowered ballistic trajectories of spaceflight. Fusion is also indispensable for power to the colonies. Since we can not carry vast quantities of manufactured articles or food from Earth to Mars, we must have tools specifically qualified to produce needed materials and articles from the raw materials of that planet. This requires not only very high quantities of power per capita, but also energy flux densities at least four times those prevailing in U.S. production. We require a universal class of tools, to use such very high energy flux densities; we require the self-focusing characteristics of lasers and particle beams, for example, which enable us to conquer every problem of materials. To feed the colonies, and long-range manned interplanetary expeditions, we require not merely present biotechnology, but the more profound capabilities locked up within optical biophysics.

It should follow, that if we can create and maintain viable cities in artificial environments on Mars, the Sahara and the Gobi deserts ought to be mastered easily by using the same technologies on Earth.

In the condition of mankind estimated to be most primitive, so-called hunting and gathering society, the potential human population is in the order of a maximum of about ten million individuals. Today, we are approaching five billion, the greatest portion of which increase has been the product of the Golden Renaissance and Colbert's, Huygens', and Leibniz's launching of the modern scientific-industrial revolution. There are limits to natural resources for the lower beasts, but not for mankind's technological progress. We have moved upward in potential population-density, through the maritime fishing revolution, the ensuing agricultural revolution, and so on. Today, by increasing the energy flux density of modes of production by a factor of about four, which these new technologies will permit us to do, we effect a revolution in the meaning of the terms "primary materials" and "natural resources." With sufficient abundance of cheap and coherently organized energy per capita, at sufficiently great energy flux densities, there are no limits to natural resources for mankind anywhere in this universe. With mastery of optical biophysics as well, the frontiers for mankind become immediately implicitly limitless.

The measure of effectiveness of weapons in particular, and military forces in general, is firepower and mobility. In production, firepower and mobility is called productivity. Both of these equivalent qualities are reflections of the technologies embodied in the construction of those scientific instruments which we call capital goods. It is impossible to introduce through capital goods, those new technologies which increase the productive powers of labor, without implicitly creating the means of production of weapons of comparably increased firepower and mobility. It is impossible to introduce the capital goods needed to produce weapons of increased firepower and mobility, without creating thus the productive capacity for effecting comparable increases in productivity.

That is key to effective "crash programs." By accelerating the use of new technologies for defensive weapons, we create the new technologies of production for accelerating productivity in the economy generally. The latter aspect of the process upshifts the economic process, to the effect that a large-scale "crash program," so directed, costs the nation not a single added net penny. The gains in productivity produce marginal increases in per capita output in excess of the military expenditures which foster those gains in productivity.

These gains in productivity proceed in nonlinear jumps, as I have indicated. Thus, it would not cost the United States a single net penny to construct a colony on the Moon beginning some time during the next decade, nor to work toward building a colony on Mars by approximately thirty years ahead. The spillovers of increased productivity into the economy as a whole would vastly more than pay for the research and deployment of such colonization projects; probably the payback would be approximately tenfold. The "systems analysts" might argue that their computers tell them this is not possible; but as long as they cling to the delusion that "cost-benefit analysis" can be based upon systems of linear inequalities, they are ignorant of the fact, that technology-driven economic processes are nonlinear.

Science-Driver 'Crash Programs'

From the foregoing and related considerations, I propose the following to be at least approximately true:

1) That all science-driver "crash programs" must have a mission-orientation, based upon a task which subsumes required solutions to each class of the problems to be solved as a by-product of that undertaking.

2) That the mission orientation of all science-driver programs, must encompass the full spectrum of all frontier developments in science, and that the work must be organized on a properly defined conception of the geometrical composition of physical space-time.

3) That no boundaries be constructed among fundamental research, development of prototypes and generalized production and deployment of new technologies.

4) That the method of development must be in the best tradition of the scientist devising materials and instruments in collaboration with manufacturers and tool-makers. All producing facilities included in the spectrum of required classes of production of materials and instruments, must assign corners of their facilities to work with scientific

teams in the same mode scientists work traditionally with tool-makers in the construction of scientific instruments. In short, the extension in scale of the normal practice of good scientific research, into the sphere of generalized production as such.

5) That no restrictions on area of fundamental research be assigned. Any research which bears upon any class of problems subsumed by the primary mission-assignment, is implicitly authorized research.

It would be an error, if the task-orientation of the SDI were limited to a list of projected military requirements. The proper mission orientation adopted as the mandate of the program should be the Moon-Mars colonization task. Each weapon system developed, should be developed by accelerating the by-products of the primary mission assignment.

The point is, that the various of the scientific capabilities for devising the military and capital goods products required, must have a common coherent basis. This basis must be defined by a task which explicitly subsumes all of the relevant technologies, and which taxes to the limit the foreseeable potentialities of each and all of those technologies.

Let us proceed to colonize the Moon and Mars, as Krafft Ehricke committed himself to implementation of this process. Along the way, we have a military problem to solve, which the technologies of space colonization are best suited to solve. Being patriots and world citizens, we shall solve that intervening task, but we shall solve it best by never taking our eyes away from our primary mission assignment. Once civilization is secured, and the productivity of labor throughout this planet increased greatly by the technological revolution flowing through our SDI task, we shall have established the more powerful economy we require to begin actually the colonization, first of the Moon, and then of Mars. All this we shall do best, if we view the practical task of colonization of Mars as a necessary way of bringing to all of mankind a vision of as man in the universe, and thus fostering the opening of the long-awaited Age of Reason.

That search in conquest of space, for this higher moral condition of mankind, is the great scientific and moral legacy of the Oberth group and its crash program exertions. We need not put that legacy aside because of military needs; we shall solve the military tasks best, if the light of the stars is never out of our eyes. Let us pledge to the memory of our beloved Krafft Ehricke, that we shall never do otherwise.

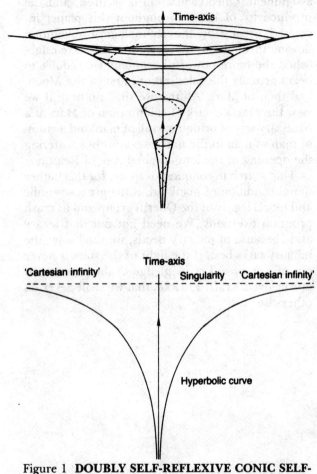

Figure 1 **DOUBLY SELF-REFLEXIVE CONIC SELF-SIMILAR ACTION.** An economic process where growth of productivity is caused by consistent technological progress under conditions of rising energy intensity and capital intensity can be described as the compounding of conic self-similar spiral action with conic self-similar spiral action (a). The result is a hyperboloid.

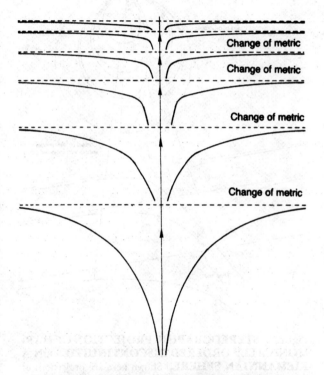

Figure 2 **HARMONICALLY ORDERED SERIES OF DISCONTINUITIES IN TRIPLY SELF-REFLEXIVE, SELF-SIMILAR CONICAL ACTION.** Increasing the energy of the system means that at the point of apparent discontinuity, the continuing action occurs on the surface of a larger concentric sphere. There is a change of metric.

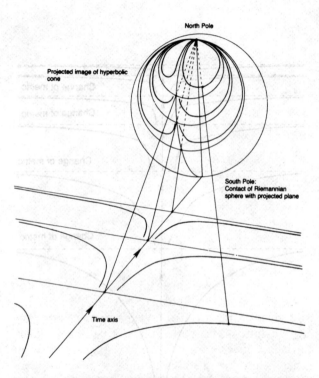

North Pole

Projected image of hyperbolic cone

South Pole:
Contact of Riemannian sphere with projected plane

Time axis

Figure 3 **STEREOGRAPHIC PROJECTION OF HAR-MONICALLY ORDERED DISCONTINUITIES ON A RIEMANNIAN SPHERE.** Shown here are projections of the action in Figure 2 onto one sphere. The so-called Cartesian infinities are now removed.

**VICE-ADMIRAL
KARL ADOLF ZENKER**

The Current Danger of Appeasement

During our long flight from Frankfurt to Washington yesterday, I was asked by a gentleman of our party that I, being one of the oldest members of the Schiller Institute in West Germany, should say a few words to you this morning.

So at first I give a hearty welcome to all the participants of the Krafft Ehricke Memorial Conference on behalf of my colleagues in Germany. After having heard the splendid speeches which have just been given to us, and which already went deeply into the material, I am now rather hesitant to continue. But then as an old maritime man, and an old naval officer who takes an interest in political developments in our time, I would also like to draw your attention to the fact that we are now living in a very critical period, as concerns the freedom of the world. I feel that our alliance is losing the ability to prevent the Soviet Union from achieving its dangerous and ambitious goals. For many years the Western powers went from conference to conference, as in Geneva, Helsinki, and Stockholm, and tried to persuade the Soviet delegates that even the Soviet Union and its satellites must play a positive role in maintaining peace and human rights, and that all nations should come to a controlled disarmament, and to a total condemnation of nuclear war.

But while the Western powers really diminished their military potentials, the Soviets gave only nearly empty promises. And in fact, they built up their forces in a tremendous way, in the army, in the air force, in the navy. In this way they are trying to get superiority over the West, and to reach their goals without necessarily going to a hot war. It seems that by means of the above conferences, they wished to gain time by making the politicians of the West believe that there is nothing to be afraid of.

What I feel to be so dangerous in these days are the tendencies, seen in the United States as well as in my country, to be resigned, and to do nothing but try to live on good terms with the East bloc. We know this tendency by the speeches of a number of politicians, high-ranking officials, and even managers of industries and economy, and day by day, by radio, television, and newspapers. Should these tendencies become the deeply felt opinion of the majority in our nations, there would be severe consequences. This development would weaken the decisiveness to defend our freedom, and would endanger our alliance. Inevitably the United States and Europe would begin to drift away from each other. But none of our countries is able to preserve its freedom without the help of the alliance. The Soviets would achieve their goals without firing a single shot.

We are called upon to avoid this with all means at our disposal, and every one of us has to play his active part in doing so. I hope, and after all I have already heard, I am sure that this conference will show us ways we are able to go.

GEN. WILHELM KUNTNER

The Soviet Decoupling Drive

Thank you very much for the invitation. I am in the same boat as the vice admiral, not in the German navy, but I got the same surprise this morning, being told that I should prepare a brief address for you. Therefore I am not prepared for a long speech. It's always a danger as long as I am standing here, that it will be longer than I was told it should be. But I am a trained soldier, and I will try to behave my best.

First of all, it has already been said that I come from a neutral country, and I have already had opportunities to talk about the imbalance between the two military alliances, and therefore it's not necessary for me to say anthing more than was already said. It is correct, and it was my first idea when I had the first task some years ago to try to give an objective view. I think this was the idea— to get an expert from a neutral country, which has

59

the advantage of an objective standpoint on these problems.

But very soon I found out, as a human being living in the Free World, living in a pluralistic country within a time I would call a cold peace, with the danger of war—I would say a pseudo-religious war—it is very difficult to be neutral as an individual. Neutrality is an institution which was born in the last century, more or less when we were confronted with imperialistic wars, and things like this. Perhaps it was easier for the individuals in the neutral countries to stay out completely, and have something like a psychological neutrality as well. This is not possible any longer. It is an official view of our authorities as well that we have to stay outside for historical reasons. For historical reasons we have to stay out of military alliances, but there is no obligation in neutrality as we understand it, for a psychological, or mental, neutrality for the individual people in the country.

This gives me the right to speak to a problem which more or less starts with the military alliance. I would try to go a little bit further. I think that what we need in the future is, I would say, the ideological dreaming of the alliance, of all those people in the world who believe in freedom and human rights—which is very much our way of life and our quality of life—against a system which has no belief in all this.

Therefore I think, and I quite agree, that we are getting into a very dangerous situation. And this dangerous situation has more extensive background, as I already heard today. It is the problem of all the Eastern European leadership, that they are confronted with many, many domestic problems, and therefore they are seeking for some success. I quite agree with what Helga LaRouche said about the Soviets' political objectives, but they only have a chance to reach them if they work out from a military standpoint. Otherwise they wouldn't work. This is one thing.

Let's just look at the Soviet Union—and I have had a chance to talk about this at more length on some previous occasions—at the frictions and their industrial work, their agricultural problems. One of the last miracles you may remember, perhaps, in the last 20 or 30 years, is that everything was discovered by the Soviets. And one of the most recent is that they sow their grain in Kazakhstan, and the harvest is done in Texas! And there is for everyone—and I am probably one of the very few in this room who knows these countries, and who knows these people in these countries since I live very near them—there is a decline in the population's belief in their leadership. And there is another problem, the increase of the Asian population and the static situation in the European side of the Soviet Union. All this might, and will, create in the future, problems for the leadership. To overcome these problems, one possibility is for them to achieve success.

And success would be, of course, to decouple Europe from the United States—to invade mentally at least, to subvert pluralistic countries in Europe. Whether they are part of alliances, or whether they are neutral, wouldn't matter very much for them, as long the countries are in the category of the so-called capitalistic and imperialistic world. Therefore they are looking for this success, and this is the dangerous thing.

Let me give you my two last ideas, and then I think I have overdone my 10 minutes already. We shouldn't get too worried about what we see on television, or read in the newspapers, about peace movements in this country, or peace movements in Europe, most of whom are minorities, being a mixture of sympathizers, useful idiots, and the avant-garde of Marxism-Leninism. We shouldn't get too worried, but we should think: In the Free World they have the right to ask for peace, whatever they understand by "peace." And they can run around. And I would say, in our Free World, whoever is not looking for peace should be sent into prison. On the other hand, if you look at the Soviet Union, or their partner countries, if anyone there asks for peace and freedom, then he is already in prison. This is the big difference we should think about.

The final thing I want to say is that I ask everyone not to get disturbed too much by all these demonstrations and movements. I think we should not let minorities and small organizations kidnap our love for peace and freedom. This is a very important thing. If we go around like this, then we are standing in the corner, always being the cold warriors. We want peace, but we want it in freedom and with human rights. Therefore no kidnapping on this!

The last thing: We should get away from being afraid of disturbances between Western Europeans and the United States. This is an idea which is injected from other sources into our societies. The Europeans are very well advised to think about this new strategy of survival and to cooperate in this. Otherwise the time might come when they stand out in the rain without an umbrella, and when they are reduced to the living standards of an undeveloped country. I think some of the governments have understood this, the population understands this, but some of the politicians don't want to understand it. But with politicians you have problems. Most of them are like a cork which wants to swim on any liquid, wants to look out very high, and has very little depth.

Having had a chance today not to speak after Demosthenes, I still would like to excuse my bad English, and I must repeat myself: I have a problem with this language—I have the same problem which I have with my wife. I love her, but I have no control of her. Thank you.

VINCENZO CAROLLO

Europe, America, and the Soviet Threat

It is known to everyone that the great problems dramatically experienced by peoples today are those of peace and of the economic order for development and justice in the world. These two great problems are not detached and alien from each other, but one conditions the other, and the consequence is that either they be resolved contextually and harmonically, or they will continue to feed threats, divisions, and death-plans between the opposite political alignments of the current world. But what international political alignment sincerely wants peace in the service of the development of peoples, and the development of peoples at the service of peace, beyond illusory, disguised, and deceitful tactics?

In *Left-Wing Communism: An Infantile Disorder*, Lenin decreed: "One can conquer a more powerful enemy only with the maximum exertion of forces

and under the necessary condition of the most dil-
igent, accurate, attentive, and skillful utilization of
even the slightest split among the enemies, of every
conflict of interests between the bourgeoisie of the
various countries, among the various groups and
the various species of the bourgeoisie within each
single country, and also of every slightest possibility
of gaining an ally, however temporary, uncertain,
inconstant, and unfaithful."

In short, the Leninist "gospel" sanctions the
adoption of a tactic of "manipulation, deals, and
compromises" with whomever lends himself more
or less consciously to "facilitate, accelerate, consol-
idate, and reinforce" (in Lenin's words) Bolshevism
in the world.

The history of the past 40 years confirms that
these doctrinal principles are concretely operative.

This version of history teaches us that a war
fought for the expansion of communism in the
world, however gradual, is not a war: It is the "liberation of peoples" and hence a condition of peace:
It is peace.

When a people decides to defend itself in the
name of its independence, sovereignty, and liberty
from some communist attack, direct or indirect,
well, in that case, that and that alone would be
responsible for war.

Belgium, France, England, and Holland were
judged as warmongers and imperialists when they
dared to defend themselves from the invasion of
Nazi Germany as long as the latter was an ally of
the Soviet state; the socialist parties of France and
England were considered opportunist and traitors
the moment they did not oppose the sending of

military aid to Finland when it was invaded by the Red Army in 1940.

But when in 1941 Hitler declared war on Russia, the Western powers which had been judged a year earlier as warmongers and imperialistic by the communist power, became immediately, and provisionally, democratic.

With absolute ideological consistency, Soviet Russia right after the end of the Second World War transformed the "zones of influence" recognized for it in Eastern Europe by the Yalta and Potsdam accords, into zones of direct "sovereignty"; having promoted in 1947 the civil wars of conquest in Greece and Turkey, it gave orders to Yugoslavia to invade the northern region of Italy, Istria; despite the American withdrawal from South Korea, the Soviet armies remained in North Korea, rejecting the free and democratic vote for unification of Korea; it started the war in Vietnam, and attempted, also with force, to attract India to itself; it occupied northern Iran, despite the fact that American and British troops had departed.

Cuba became a base of military occupation in Central America, as Libya was to assume the same role in Africa and the Mediterranean; Angola, Ethiopia, and Madagascar, and Nasser's Egypt marked significant steps in the so-called missionary expansionism of the Soviets in the world.

At the beginning of the 1960s, the U.S.S.R. started to produce intercontinental ballistic missiles (ICBMs), with the goal of overcoming its military inferiority with respect to the United States, which did not take advantage of its own superiority to set back or block Soviet expansionism in the world.

In 1972 SALT I was signed and in 1979 SALT II; but the U.S.S.R. never really respected the conditions and pledges in the treaties.

The Fraud of Peaceful Coexistence

America started to theorize about the policy of peaceful coexistence; it proposed to construct peace in the world by means of economic and financial aid; but for the leading state of the Communist International, peaceful coexistence must be understood as the refusal of any form of Western resistance to Soviet expansionism, and economic and financial aid should be rejected if it automatically leads to the creation of direct relations between the recipient countries which were not yet definitively communist, and the Western powers.

And hence we have the duty to ask ourselves: "If this is the reality which faces the states of the Western world, what should these Western states do, in order not to fall into the Soviet sphere of influence or sovereignty?"

On this matter, the democratic West shows itself uncertain, contradictory, and in certain aspects weak—weak in the sense that it is afraid of organizing itself as a coherent and unified force from the political and economic standpoint.

Certainly, the democracies that rule the independence, sovereignty, and freedom of the Western states are building their own representative force in the midst of the free consensus of the populations.

However, no citizen is forced to give his consensus to those who offer war.

Every political party therefore studies the way

to guarantee the peace, even at the cost of always conceding something to the power which by "peace" means the expansionistic war of communism in the world. And thus for some, the steps toward peace become a gradual, albeit painted-over, scaling down of one's own power; for the Soviet empire it is instead a missionary-like, constant advance toward greater power.

Since 1971, for example, the *Ostpolitik* was supposed to reinforce the Soviet presence in Eastern Europe on the economic policy level, as the price the West agreed to pay for the tranquility of Western Europe.

The so-called Second International has been studying what could be offered to Russia in addition to that which it has already obtained and conquered. In this way they convinced themselves of the illusion that they were creating new conditions for peace, pretending, however, that they didn't understand that for the "internationalism" of the Soviet leading state, even the slightest result obtained is always a step toward the path of further expansion, and not its definitive conclusion.

When between 1968 and 1975, the American allocations for defense were cut by $40 billion, did Russia do the same?

When, beginning in 1977, President Carter suppressed the B-1 bomber, delayed the production of cruise and MX missiles, suspended the production of Minuteman missiles, and suppressed the neutron bomb, did Russia perhaps lighten up its military budget or did it instead continue to produce new atomic weapons, to the point of reaching nuclear superiority over America itself?

And yet in those very same years, the heads of government of the major European countries maintained that they had only one duty to fulfill: that of mediating between the so-called intransigence of Nixon and Soviet aggression.

Mediatory visits to Moscow multiplied, by Macmillan, Wilson, de Gaulle, and Brandt.

What were the concrete and definitive results?

I believe that it is not difficult to understand their nature.

The Problem of Western Europe

But today, and more precisely in these years of the 1980s, what is the global correlation of forces, and what can be predicted?

I maintain that today the true international political problem is represented by the role of Western Europe.

From the strategic standpoint, Europe is geographically essential, but from the political standpoint its position unfortunately lacks organic cohesion.

Within Europe some powers are carrying out disunified and competitive activity, which fatally weakens its unified character. In some political and cultural circles, the conviction is spreading that the neutralism of Western Europe can be more fruitful than Atlantic consolidation.

It is a fact that about one month ago, it was decided to standardize the conventional armed forces within the European Atlantic states, but it is also true that this decision carries with it an apparent justification of blocking strategic or "space" weapons—in conformity, obviously, with the essential demands of Russia.

To understand better the present attitude of some European powers vis-à-vis the beam weapons project, it would be useful to recall the refusal of France and Britain in 1961-1962—when U.S. President J.F. Kennedy proposed the constitution of a multilateral force (MLF)—to be placed under a collective contract by the participating states. The Carolingian conception of the time refused to accept the principle of the collective contract, in the fear that it could become preeminent control by the United States.

This Carolingian conception of the relations of some European states toward the other states and toward America, itself has not died out. That explains many things.

As long as Europe remains a sum of states each interested in its own respective power and not the unified power of Europe as a whole, then it will be little by little, inexorably, emptied out.

The first evident target will not be continental Europe, but the Mediterranean. It is absolutely certain that the partial or total control of the Mediterranean or its strategic paralysis, is a real and inalterable objective of the Soviets.

Malta continues to be the supply depot for Soviet ships; Libya continues to be the base of Soviet military expansion in Africa and the Middle East; Syria continues to maintain, on behalf of the same Russia, the ferment of disorder and death in Lebanon.

And then what use is Greece's neutralism, the apparently self-sufficient exclusivism of France, the search for new lovers on the part of West Germany?

This brings us to the second aspect of the fundamental problem which today engages all the Western powers: economic relations with the East.

It is not insignificant that just a few days ago, the East bloc countries gave the EEC that recognition which for decades they had consistently refused: They proposed to set up direct relations between the EEC and the Comecon.

Gorbachov has understood that the European states have productive structures which are greater than the levels of their internal consumption, and that it is not easy for them to conquer new markets in the Western world, given that the industrial sector has the same problems as Europe and the underdeveloped sector lacks indigenous economic resources to guarantee strong consumer markets.

Therefore, to penetrate into the economic reality of Western Europe offers Russia today multiple occasions for political blackmail.

For Leninist Russia, as I recalled, any economic relation with other different states is not and cannot be an end in itself. But for the last several years, commercial and financial relations between Western Europe and Russia have broadened and, following the same road, they will increase more.

Russia supplies industrial and agricultural raw materials; Europe supplies manufactured goods and high-technology products. Why all of this?

Will the Alliance Survive?

We have to honestly keep in mind that the blame must be placed not only on a disorganic Europe afflicted by internal rivalries, but also on the United States. The U.S.A. has pursued, from 1979 onward, an economic, financial and monetary policy which has been twisting and turning, episodic, and not always in harmony and solidarity with the European countries.

The fits and starts of the dollar in the international field, where overvaluation and undervaluation are not derived from objective laws of economics, but from arbitrary actions by apparently autonomous political powers, they have not created order in international and European relations.

The recognition of the ECU [European Currency Unit] as currency of reference and accounting in the international field has not been good for anyone. It has not been good for Europe, and it certainly has not been good for the economic order of the West.

To maintain today that rule over the economy should be led by Japan, the United States, and Europe is no doubt positive from the conceptual and political standpoint.

But everything will also fail if the three great power blocs were to remain competitively distanced from each other. Instead we must install a harmonious solidarity which takes into account the reciprocal, recurrent, seasonal difficulties of each of the three blocs.

If this is not considered and is not accepted, then it would be a bit like calling "healthy" an organism in which the heart, lungs, and liver function each on their own. In this case the organism as a whole will definitely be the one to suffer. The heart, which could not feed on the red cells of its own organism, since the lungs and liver did not function in harmony with it, would decide to have its blood injected from outside—in our case, from Eastern Europe. In this case the blood from the East would have ingredients very different from those in the Western blood.

Economic reality is not less relevant than strategic reality; the economy cannot help being an integral and vital part of the overall Western world strategy.

It is true that the Third World should be aided and developed; but it would be shooting ourselves in the foot to create the conditions for Europe to become a Third World over time.

Facing bitter prospects of this sort, imagine the fatal situation that would prevail in the next decade, if not only will methane gas flow from Siberia to Europe, but also major investments of ever more substantial financial and economic resources.

Then new conditions will be created, so that the economic relation becomes a permanent political relation.

Who wants this? In words, no one does, but in fact there are well-founded suspicions that some powers desire it, even if such desires are from time to time powdered and painted over.

Krafft Ehricke, Peenemünde, and the Scientific Origins of Space Technology

Afternoon Panel
June 15, 1985

HELGA ZEPP-LAROUCHE

In Honor of a Cultural Optimist

I want to start the afternoon panel, which will be opened with a selection of film material, both from the famous movie *The Woman in the Moon*—which can be called one of the starting points of space travel because it recruited a lot of young people— as well as of some of the early experiments and recent experiments, because most people just don't have an image of what it actually means to conquer space. This film material is very pedagogically usable, and we are planning to replicate it in video form, so that it can be used as educational material throughout the world, not only in the United States, but in Europe and in the developing countries, so that it captures the imagination of people.

Given the fact that this panel is in the honor of Krafft Ehricke and the history of space travel and rocket technology, I want to say a couple of words about Krafft Ehricke himself, as I knew him. I think

the thing that impressed me most about Krafft was
that he was one of the most exciting cultural op-
timists I have ever met. Even in the late phase of
his life, when he had a very difficult combination
of different cancers, and he had to undergo different kinds of very horrible treatments, he did not
lose hope once. He was still working on his books
up to the last moment, and he mediated this tremendous hope in mankind, which is typical for
humanists of his tradition. This is not a surprise
because he was educated, as he has emphasized in
his many writings many times, in the German classical tradition of Schiller—the Humboldt, Beetho-
ven, Mozart tradition.

When I called him the last time for a long phone
conversation in October 1984, which was when he
last came to Washington to receive an honor from
the space community, it was very moving, because
he said: Look, I have recognized at the end of my
life—and he was quite aware that he did not have
very long to live—that technological development
is an absolute necessity, but we have to concentrate
at the same time on the development of the character. The perfection of the human being and the
perfection of the soul, in the way Schiller has described it, is absolutely necessary. When you see
this debate we have today about whether progress
is good, if you see the Greens coming up as a neo-
Nazi movement in many European countries, there
is a lesson to be learned, and that is that technology
is a must, an absolute evolutionary necessity, as
Krafft said, but it has to be combined with the en-
noblement of the soul. The two have to go hand
in hand.

One thing about Krafft Ehricke is that he, quite independently from us—in a certain sense our routes crossed relatively late, five or six years ago—came up with a theory of evolution, with a conception of negentropic development of the universe, quite on his own. For him the evolution of astronautics was the only way in which man could overcome his limitations. He understood that evolution is not a choice, it is an absolute necessity, and that whenever man violates the need to negentropically develop, he violates the laws of the universe at his own risk.

Krafft had a fundamental belief in the creative ability of man. But he thought that there was something necessary to lift people out of the mud of quarrels on Earth, which he called the "extraterrestrial imperative." It was the idea that man would only change his character in the larger context of the cosmos at large. He said: Take, for example, my own forced departure from Europe. Having grown up in Europe, he had a classical European education. Still, he said, my character changed when I came to America, because in America there is a different type of people. They are human beings like Europeans, but they are in a certain way more pushy, more tough, and he came to the conclusion that the mixture of the old European tradition and culture, and the sometimes a little bit more couragous American way of doing things, combined, would add something to human nature—which happens to be an experience I have had myself. He said, that through this extraterrestrial imperative, man would become different.

Living in Space

Having the experience of living in modularized space stations in space, on the Moon, or on Mars, where man is completely responsible for any conditions he creates, where there is no biosphere, you can't walk out of the door, there is no nature which provides everything for you—this will change man. If you go to the Moon or to Mars, man is completely responsible for all the conditions he is creating, which obviously means people will have to become more rational. On the Moon, if you are an anarchist, and you leave your modularized environment, well, you are in trouble. He developed this philosophy, and I think it is a very deep philosophy, that man would become mature, and the true dignity of man, of which the European humanists were dreaming two hundred years ago, would be realized in space. From that standpoint, for him the Moon was just a door, a gate, which would be opened, and he called the colonization of the Moon the cradle of space civilization, and from there we can go beyond. I should also say that he had a very clear idea of what civilization means, namely an overcoming of brutality.

This is something worth thinking about, because our world right now is characterized by a lot of brutality. When we say that Western civilization, or human civilization, is about to be lost, what we mean by this is that we are about to be taken over once again by the brutality which characterized previous ages, and obviously we have to go in the other direction. To reflect some of the discussions we had during lunch with some people, the question came up, that since there are so many hungry people in Africa, Latin America, and Asia—and even in the United States there are 50 million poor people—why go into space? Is this not a total waste? I think it will be one of the most important accomplishments of this conference, for it to become clear in the discussion, that for economic and scientific reasons, there is no way in which we can develop the poor countries of this Earth without going into the next technological revolution.

Then there's all this talk about overpopulation, which we know is a Rockefeller baby, or otherwise known as a fraud. If you look at these countries in Africa, they are underpopulated. Sudan, a country which is nearly as big as Europe as a whole, has about 20 million people. Belgium has more. It's not overpopulated, obviously. Hermann Oberth, for example, as a by-product of his projects, first had the idea, which then was elaborated by Krafft Ehricke, of placing huge mirrors in space and capturing some of the sunbeams, and redirecting the Sun's energy in a very concentrated way to icy areas of the world. This way you could melt the ice in the north, you could make habitable areas for many more people, and you could improve agriculture. You could prevent crops from being destroyed by frosts, you could correct the weather in a very scientific way. There are many, many uses out of this without any question.

This should be a subject of discussion: There is no way in which the hunger and poverty on this Earth can be overcome without that leap in technology. The social work approach may have its uses in preventing people from going into the pit. What this person from Philadelphia was saying [in a comment this morning], was very beautiful; it's very important, that there are people who are doing this, because otherwise, people would become prostitutes and pimps, the drugpushers would take over, etc. But you cannot save mankind in this way anymore; it's too late. You need a dramatic big change, the technological revolution in a big way, and that can only come through in the same way as the steam engine and nuclear energy. You have to make a jump, a revolution in technology. What is the only way we can get these cowardly military, who right now let themselves be taken over by the Eastern Establishment, to move? How do you get some steam into them? Obviously, you need those kinds of projects we are going to show to you right now, the idea of colonization of Mars, that there are no limits to the human mind, that the human mind can develop beyond every border possible. That is precisely the kind of imagination which people have to have in order to progress.

One last word on Krafft Ehricke himself. If you think about how mankind has progressed from the Stone Age, really, if you think back all these years, tens of thousands of years, really qualitative breakthroughs have been made by very, very few people. Once in a while, because of a certain tradition, our tradition, the Judeo-Christian humanist European tradition, and the way that culture reflects itself in other areas, like Asia, Latin America, Africa, etc., we have made progress for mankind. But within that tradition, it was due to very few people, who, because they committed their lives to tremendous work—and as Schiller said at the end of his life, genius is work—that progress was made. I would say that Krafft Ehricke absolutely deserves the way people remember him, as one of those individials, who represented a singularity, which brought mankind a qualitative step further. And I think we cannot honor him in a better way than trying to be like him.

Frau im Mond, Fritz Lang's 1929 film, was produced with the scientific consultation of Hermann Oberth, who is shown here (second from left) with Fritz Lang (third from left) on the set for the film.

KONRAD DANNENBERG

Peenemünde and the Scientific Origins of Space Technology

I am certainly very glad to be here, particularly after you have all seen what happened in the early stages of rocket development. When I first met Krafft Ehricke in Peenemünde, I was really impressed by the knowledge, he already had at the time. He certainly had read Professor Oberth's book, *Rocket to Interplanetary Space,* which was published for the first time in 1923, and where Oberth told all these people what they should have done. They should have known better than to make all these tests you just saw [referring to a movie of various early rocket tests]. Ehricke was familiar with the mathematics; he was well versed in all kinds of mathematics activities, and many of the things he had to do later on were strongly mathematically oriented. But let me go back and show you some slides of the old days in Peenemünde.

This is a V-2 (Figure 1), which you have seen already in the movies we just showed. The big question for this audience here may be, why is the V-2 really so important? After all, we have seen that people had built rockets before. In fact, rockets had been used for warfare for hundreds of years. The Chinese had been doing it, it can be documented, since 1232; the British and the Danes had rockets in their artillery batallions, which were used to bombard cities. So what is really the big step with the V-2?

I would like to summarize the answer briefly, because I think it is very important for all the follow-up discussions. When Wernher von Braun finally convinced the German army to put some money into the thing, in order to build bigger rockets than we had seen up to then, he basically made two decisions. First, he wanted to go to liquid propellants, because most of the mishaps we had seen before were due to the fact that solid propellants were used in most of these instances. So, he decided for that reason not to work with solid propellants. Solids also have the disadvantage that they are not only very dangerous, but also very difficult to control, particularly with their cut-off conditions. Mainly for that reason, von Braun decided to go to liquid propellants. Of course, he also had read Professor Oberth's book, so he knew that hydrogen-oxygen was the best propellant combination available. This is still so today; we can always build better liquid rockets than we can build solid propellant rockets.

Let me show you where we stood at the time Dr.

von Braun started his A-4/V-2-activities. Here is a picture of Prof. Robert Goddard in this country (Figure 2), and you see his rocket launch structure, which is typical of the type of things that went on not only in Germany but all over Europe at the same time. Many of the people like Professor Goddard had drawn the conclusion that the real way to go was to use liquids. Now, Goddard was not as ambitious as Dr. von Braun; he did not want to build, in a way, artillery bullets that could fly over very large distances; he only wanted to reach higher altitudes. That was his aim, and he decided that even just to obtain great altitudes, we should really use liquid rockets. The rocket chamber is all the way on top of his configuration; the combustion chamber is in front of the nozzle. He decided, since he did not have a good guidance and control system at that time, to drag his tanks behind the rocket motor.

Below in the photo are the tanks where Goddard kept his propellants; the propellants were fed from the tanks through the tubes on the side, because they were now heated by the rocket exhaust. Under pressure generated from the heating of the propellants in the tanks, they are pushed out and into the rocket engine. This was still a very primitive thing, since Goddard at this time did not have guidance and control systems (which he developed later).

But Goddard's greatest shortcoming was—unfortunately, for this country—that he did not get federal support. I am sure that if he had had the funds that finally were made available in Peenemünde for the construction and the production of the V-2, he probably would have come to very similar conclusions as von Braun. That was von Braun's big advantage, that he could convince the military, and you saw earlier in the film that the military watched very often and very carefully these rocket tests, so they could be convinced of some benefit in the use of rockets to replace artillery. Fortunately for Germany, the Treaty of Versailles after World War I had forbidden Germany to have heavy artillery, so somebody had to come up with a way of replacing heavy artillery, and that was the idea to use rockets, which was the main reason that finally von Braun developed and built the V-2.

In this aerial view of the Alabama Space and Rocket Center (Figure 3), you see the entire history of rocketry. You see the V-2 rocket, a real little one standing in front of the Saturn I, which is in the background. The tallest standing rocket is the Saturn I, and Krafft Ehricke developed engines for the second stage—I am going to make a few remarks on that later. Next to the V-2, which stands right in front of the big one, you see two or three Redstones, the slim and slender ones; the fat one is a Jupiter. Other vehicles were built after those, and all the way in the foreground, you see lying on the ground a Saturn V. So here is the entire rocket history in front of you in this one picture. I think you should consider to have one of your conferences one of these days in Huntsville, where you can see all this hardware firsthand and not just in a picture.

As I said earlier, the V-2 was built as a military weapon, and you see what the basic idea was on how to deploy it (Figure 4): It could be assembled in the little shed in the back, and either on a relatively simple road transporter, or, in this case, on two dollies running on small gauge railroad tracks pulled by a locomotive. It could be pulled out, it could be erected at the launch site, and it could be launched without too much effort. The V-2 was empty at this time. We had to keep it empty until we had directed it, and the filling of the tanks could take place only after it was in its launch position.

You see here (Figure 5) the picture we saw already in the film. This is a V-2; it must have been a launching in about 1941. You see the people had to get to the guidance and control system on top of the vehicle, and they also had to do quite a bit of servicing down at the power plant.

Peenemünde

Peenemünde has been mentioned here repeatedly, so I will show you where Peenemünde is located (Figure 6). In the lower part of the map is the northern part of Germany. All the way to the west is the former state of Mecklenburg; Poland now extends all the way to two-thirds of this picture here. Peenemünde is very advantageously located on a cape on the island of Usedom; you see here all the potential trajectories emanating from there. You can see why Peenemünde was picked; its location had several unique advantages. First, the area was sparsely populated, so the land was still cheap, and people would not be endangered by the launching of these vehicles. Second, you could launch the vehicles along the coast of Germany at that time, and you could obtain a range of about 200 to 300 kilometers, so you could also observe the impact. Through this part of Mecklenburg flows a relatively

small river called the Peene, which pours its waters
into the Baltic Sea, and right at the mouth of that
river you find Peenemünde. On the other side, now
Polish, is Swinemünde; there is a similar smaller
river, the Swine, and between these two rivers, the
Peene and the Swine, there is the island of Usedom.

Here is a more detailed picture of Peenemünde
itself (Figure 7). The teststand we saw earlier, from
where the missiles were fired, is all the way on top
of the map. This was called the *Prüfstand* No. 7. A
little bit farther south was one of the first test stands
we had; therefore it was called *Prüfstand* No. 1. Still
a little bit farther south were the assembly facilities,
so you had a relatively short road from the assembly
facility to the test stands and there was quite a variety of test stands. I should also point out, a little
bit farther to the west from the Peenemünde test
station there was an air force test station. The black
spot you see almost on top there is a launch site
for a V-1. Later, I am going to show you a picture
of a V-1—a V-1 is not a rocket and you should not
confuse it with a rocket, or the V-2, in this specific
case. There was an air field located there, which
was also used to launch V-1s.

This is another photo of the Space Center at
Huntsville, and the gray body in the foreground is
a V-1 (Figure 8). In front of that is a V-2 power
plant. Behind the V-1, are some U.S. Army missiles;
the front one is a Jupiter, and behind it is a Redstone, and we will talk about those a little bit more.
The V-1 is basically an automatically operated and
controlled airplane. It has a pulse jet, which is the
tube on top of the airplane pipe structure, and

which sucks in air from the outside. That is why I said earlier it is not a rocket.

Here is a a detailed picture of the A-4 (Figure 9). This was later renamed by propaganda minister Goebbels into V-2, *Vergeltungswaffe 2,* but the technical people always referred to it as the A-4, *Aggregat 4.* There was also an A-1, A-2, A-3; and, in fact, before the A-4 was flying, we flew an A-5, which was a smaller version, about a half-scale version of this. When I had my very first contact with Krafft Ehricke, after he had come in early 1941 to Peenemünde, he had to study this picture, he had to look at all the components, and I was the fortunate person who was selected to tell him about all these things, and we talked about things as shown here (Figure 10). This is a turbo-pump, another trick that von Braun played in order to be able to build a rocket light enough to be launched.

Most of the earlier people in the films we saw did not launch the rockets going straight up because they did not have enough thrust. You need more thrust than the weight of your rocket in order to launch straight up, and in most cases people could not do it. Von Braun came up with the idea to use a turbo-pump; there is an oxydizer pump on this side, a fuel pump on the far side, and in the middle a turbine. By driving the turbine at a pretty good speed, you can put your two propellants, your liquid oxygen and your fuel—in this case 75 percent alcohol—under high pressure into your rocket engine. That is what you have to do to get a good thrust.

Here is the thrust chamber (Figure 11). These are historic slides. They were made in Peenemünde or maybe very shortly after we came over here, and that is why some of them are a little bit fuzzy. You had to fill this combustion chamber (at the top) with very high pressure propellants. The combustion chamber here operated with about 20 atmospheres of pressure on the inside, and the combustion temperature was also very high, so the walls had to be cooled. That's why the fuel was used as a coolant. You entered at the bottom of the combustion chamber, you let the fuel flow through the double wall of the combustion chamber, and then finally you let it enter all the way on the top into the individual combustion elements.

That was in fact my very first job when I came to Peenemünde. I had to develop and to design these individual combustion elements. We finally decided to use 18 of them, to generate the 25 tons of thrust we needed to lift off with a 12-ton V-2 and to get the necessary flow of propellants into the combustion chamber. All the way on top is the turbo pump, and a little bit farther to the right there is the equipment needed to drive the turbine. Of course, you have to provide power to your turbine, and it was decided to use hydrogen peroxide from a high-pressure container. The container is pressurized because then it can be made relatively small. From there, under high pressure the hydrogen peroxide is pushed through a catalyst bed, where it is decomposed into hot water steam with some oxygen in it, and that drives the turbine. Here is another look at the V-2 engine (Figure 12).

The assembly of the V-2 center sections is shown here (Figure 13). You see the two tanks here in front; the left one is a fuel tank, the somewhat smaller one the liquid oxygen tank. The fuel tank is conical; therefore, it has to be a little bit longer. The center section was a separate building element. So the tanks were not integral as they are today with all our IRBMs and ICBMs. But here, we have a separate tank and a separate shell, which consisted of two parts. These are all the things I had to show and to explain to Krafft Ehricke. I was not so much in the production facilities, so very often he also talked to other people. He also, of course, talked to all the guidance and control people in order to be able to write the first complete handbook on the A-4, the "A-4 Bible."

Fortunately for Ehricke and myself, we both worked for a very advanced-thinking fellow, who basically had the same characteristics of Dr. von Braun. This was Dr. Walter Thiel, who was in charge of all these propellant arrangements. He made the big decisions of how to cluster the individual elements, how to build the center section, and all these things. He was unfortunately killed in an air raid that took place a little later.

One of the other things Ehricke had to do after he understood how the A-4 technically worked and how to describe it, was to find out how to make it plain to the German soldier, who of course, very often was not well-educated. So Ehricke had to give all his explanations on an easily understood level. He also had to see to it that the troops, the trained soldiers of one rocket battery, could handle their complement of three V-2s, and that's why I show this photo (Figure 14). Here you see three V-2s with the erection equipment, which was attached to a transporter vehicle, a Meiller-Wagen. After

Ehricke's instructions, after he told them exactly what had to be done, three of these V-2s could be launched within two hours. That included pulling them into the launching site and erecting them; the propellants, as I said earlier, still had to be filled into the tanks. The guidance systems had to be directed. You needed the auto lights in order to know exactly in what direction you should launch your vehicle, and the range was determined by cutting your engine off at the right time. So there, the soldiers didn't have to do too much, that was done by the first automated computer on board the V-2. All these things could be done in two hours, so a rocket battery could launch three V-2s in two hours. Krafft Ehricke had a very instrumental hand in making all the arrangements and asking for the necessary equipment that would be needed in order to do this launch sequence.

Here is an aerial photo taken picture made by British intelligence (Figure 15). This picture was taken less than a month before the air raid on Peenemünde took place; and unfortunately, I have to say, that at this time, I somewhat lost track of Ehricke. After the air raid, many of the buildings of these facilities here were destroyed. Not the test stands. They are pretty sturdy, and they take a lot of beating. They were designed and built for explosions of rocket engines, so even the bombs couldn't do too much harm to them. But the buildings where the V-2s were assembled, the office buildings, were destroyed. So at that time, the whole plant moved over the entire island of Usedom, and some of the organizations even moved to the mainland. So at this time, I somewhat lost track of Krafft Ehricke.

The U.S. Army Program

Here is General Medaris (Figure 16). General Medaris was very instrumental in taking the A-4/V-2 information from the Germans and translating it into an Army program. That's why you see him shown here with the Redstone. The next rocket is the Jupiter, and after the Jupiter came the Jupiter C. The Jupiter C is just a modified Redstone, which is a little bit more powerful than the basic model.

One of the strong points Medaris had was that he recognized the shortcomings of the V-2. This is an impact area of the V-2, again a British intelligence picture (Figure 17). Down in the center of the map is London, a little bit farther up, you can see the city of Norwich. Of course, the intent was to hit downtown London with all the V-2s, but you see what actually happened. The V-2s were not completely ready for military use. In fact, the technical people were quite a bit opposed to using it at that time, but again we got a political decision: The Nazi leadership decided that it was time to use the V-2 regardless of its condition, and here you see the results.

Many of the misfirings, or not hitting the target, were due to the fact that when the V-2 returned from space, the control fins of the rocket were not too effective. A couple of V-2s came down broadside, and of course they broke up, and they were easily reflected by all kinds of other conditions. Therefore, General Medaris tried what we see here (Figure 18) for the Redstone missile, to use a power plant which is just an advanced version of the V-2 engine. And General Medaris decided for the Redstone missile to separate the warhead. It is not a two-stage vehicle; the warhead is not propelled, but now you get rid of the problem I just mentioned for the V-2. Even if it would break up and you would lose the main part of the rocket, your warhead would still reenter and hit the target. This is the first time we used this very basic procedure that cleared the way for all our missiles that are in use today.

Since the Redstone was the first vehicle that was available and the first vehicle that was also very reliable, we could also use it to build the Redstone launcher, which you see here with the Explorer satellite all the way on top, which was the first U.S. satellite in orbit (Figure 19). The Redstone, with the Mercury capsule on top of it, was also used for the launch of Commander Alan Shepard and Gus Grissom, the first U.S. astronauts in space. The Redstone is not big enough to put people into orbit, it would only take them to great altitudes of about 150 miles. . . .

At that time Krafft Ehricke decided to leave the von Braun team; he decided that it was too conventional, . . . still using the old hardware, and he thought one should switch to liquid hydrogen. That is why Krafft Ehricke left the team around this time; later he joined General Dynamics, where he worked eventually on the Centaur with the first liquid hydrogen upper stage engine.

The first satellite launch is shown here (Figure 20), the satellite rotating on top of the vehicle. The next photo shows one of the two Mercury launches

(Figure 21), and the next one (Figure 22) shows the entire series of launch vehicles available. Krafft Ehricke, who had joined the Convair division of General Dynamics, was working on No. 6, the Atlas-Centaur. He was really very instrumental in developing the Centaur, and he developed the Centaur upper stage, and he also helped develop a very unique engine. Here is an Atlas again shown at the Space Center, the one lying flat (Figure 23). One part was replaced by the Centaur upper stage, but we don't have a Centaur at the Space Center.

Here is the RL-10 engine, the first liquid hydrogen engine, which Krafft Ehricke was very instrumental in developing, together with Pratt and Whitney, which built this engine (Figure 24). You see, you don't have a double wall anymore for your combustion chamber. It consists of five individual tubes. You save a lot of weight doing it that way. You still use a turbo pump and here you see all the machinery that is part of a turbo pump system.

Here is a schematic of the turbo pump (Figure 25). I don't know exactly who came up with this idea—it's a very unusual idea—but the oxygen and the oxygen pump are at the top wall on the left-hand side; it flows through the line at the left after it has been pressurized by the pump, it goes directly to the combustion chamber, where you collect the hydrogen, which by this time has picked up so much energy, so much heat, that it is preheated and can mix with the oxygen. . . .

And here is Apollo 8 (Figure 26), which is the first manned flight to the Moon. The astronauts didn't land on the Moon in this mission, but it was the first mission when any astronaut went outside of Earth orbit, and it proved the feasibility of all the Saturn V and other systems. . . .

The Shuttle, the Space Station, and Beyond

The Shuttle engine, shown here in a schematic (Figure 27), uses a similar kind of liquid hydrogen engine, a later-generation liquid hydrogen engine from the one Krafft Ehricke developed. . . . Here (Figure 28) is an early design of a space station, similar to von Braun's. At that time, people thought the space station would have to rotate to simulate Earth's gravity. This is a current NASA design for a space station (Figure 29). You can see the Shuttle and a large piece of equipment under construction.

Finally, here are some young potential astronauts at the Alabama Space and Rocket Center (Figure 30). They are at Space Camp for a week, learning about space and spaceflight, learning that they can eventually do all these things, that they can fly up to the space station. At the Space Center, we have a space station simulator, where we simulate space station activities. One of the highlights of Space Camp is, of course, a flight inside of the cargo bay of the Space Shuttle. Of course, it's simulated; that's why we can make three or four flights a day, while NASA still needs about a month to do that.

I hope that with the Saturn V and even with the Shuttle, that these activities in spaceflight do not end. You see again the total picture here (Figure 31): The small vehicle all the way in the back is the V-2. In front of it is the Mercury Redstone, with the Mercury capsule on top of it. Behind it, you see the Saturn I that I mentioned, where Ehricke provided the Centaur engine for the second stage; that was the first big stage ever propelled by liquid hydrogen. And, of course, you see in the very front, the Saturn V.

I would also like to mention that July 1 marks three important events. It is the 40th anniversary of the "Paperclip" group's arrival in the United States, the 35th anniversary of the group's arrival at Huntsville, and the 30th anniversary of the Marshall Space Flight Center.

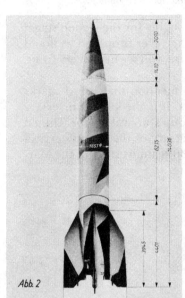

Abb. 2

Figure 1 THE V-2 ROCKET

Illustrations are courtesy of Konrad Dannenberg, the Alabama Space and Rocket Center, and NASA

Figure 2 ROBERT GODDARD AND HIS ROCKET LAUNCH STRUCTURE

Figure 3   AERIAL VIEW OF ALABAMA SPACE AND ROCKET CENTER

Figure 4 ASSEMBLY OF THE V-2 ROCKET

Figure 5 **LAUNCHING OF A V-2**

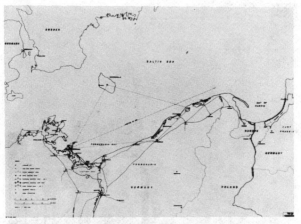

Figure 6 **LOCATION OF PEENEMÜNDE**

Figure 7 **MORE DETAILED MAP OF PEENEMÜNDE**

Figure 8 **THE V-1 AIRPLANE**

Figure 9 **DETAILED VIEW OF THE A-4**

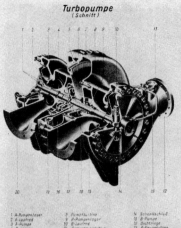

Turbopumpe
(Schnitt)

1 A-Pumpenlager	8 Dampfturbine	14 Schornsteinschluß
2 A-Laufrad	9 B-Pumpenlager	15 B-Pumpe
3 A-Pumpe	10 B-Laufrad	16 Dichtringe
4 Kupplung	11 Reglersteuerschlitz	17 B-Saugstutzen
5 Düsenkasten		18 Turbineneinlauf
6 Frischdampfgehäuse	12 B-Druckstutzen	19 Schaufelkranz
7 Abdampfgehäuse	13 B-Leckleitung	20 A-Saugstutzen

Figure 10   VON BRAUN'S TURBO-PUMP

Heizbehälter-Schnitt

1 B-Einlaufstutzen	4 Stutzen für A-Zuleitung	7 Verbrennungsraum
2 Untere Kopfkammer	5 A-Zerstäuber	8 Dehnungsnischen
3 Obere Kopfkammer	6 B-Düsen	9 B-Einlaufstücke

Figure 11   THRUST CHAMBER OF THE V-2

Figure 13   ASSEMBLY OF V-2 CENTER SECTIONS

Figure 12   THE V-2 ENGINE. The author is shown with a V-2 model at the Alabama Space and Rocket Center.

Figure 14
THREE V-2s AND THEIR ERECTION EQUIPMENT

Figure 15
BRITISH INTELLIGENCE AERIAL VIEW OF PEENEMÜNDE

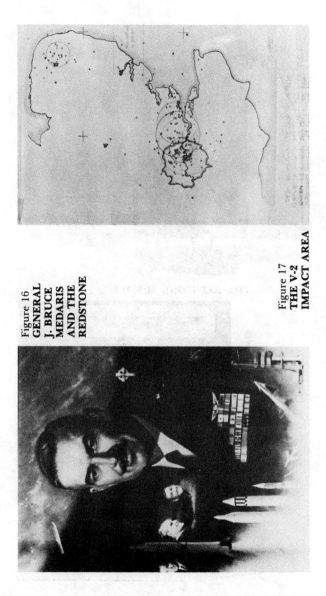

Figure 16
**GENERAL
J. BRUCE
MEDARIS
AND THE
REDSTONE**

Figure 17
**THE V-2
IMPACT AREA**

Figure 18 **THE REDSTONE MISSILE**

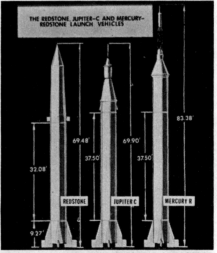

Figure 19 **THE REDSTONE LAUNCHER
WITH THE EXPLORER SATELLITE**

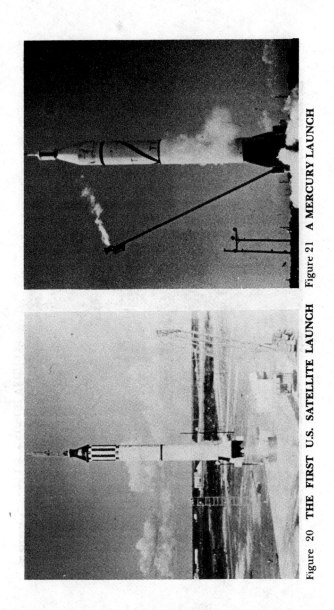

Figure 20 THE FIRST U.S. SATELLITE LAUNCH

Figure 21 A MERCURY LAUNCH

Figure 22 U.S. LAUNCH VEHICLES

Figure 23   **THE ATLAS MISSILE**

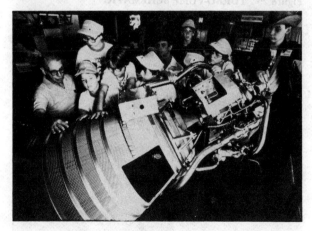

Figure 24   **THE RL10 ENGINE**

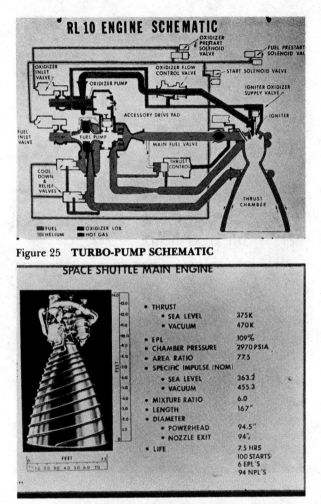

RL 10 ENGINE SCHEMATIC

OXIDIZER PRESTART SOLENOID VALVE

FUEL PRESTART SOLENOID VALVE

OXIDIZER INLET VALVE

OXIDIZER FLOW CONTROL VALVE

START SOLENOID VALVE

OXIDIZER PUMP

IGNITER OXIDIZER SUPPLY VALVE

ACCESSORY DRIVE PAD

IGNITER

FUEL INLET VALVE

FUEL PUMP

MAIN FUEL VALVE

THRUST CONTROL

COOL DOWN & RELIEF VALVES

THRUST CHAMBER

■ FUEL   ■ OXIDIZER LOX
■ HELIUM   ■ HOT GAS

Figure 25   **TURBO-PUMP SCHEMATIC**

SPACE SHUTTLE MAIN ENGINE

- THRUST
  - SEA LEVEL   375K
  - VACUUM   470K
- EPL   109%
- CHAMBER PRESSURE   2970 PSIA
- AREA RATIO   77.5
- SPECIFIC IMPULSE (NOM)
  - SEA LEVEL   363.2
  - VACUUM   455.3
- MIXTURE RATIO   6.0
- LENGTH   167"
- DIAMETER
  - POWERHEAD   94.5"
  - NOZZLE EXIT   94"
- LIFE   7.5 HRS
                   100 STARTS
                   6 EPL'S
                   94 NPL'S

FEET

-10 20 30 40 50 60 70

Figure 27   **SCHEMATIC OF THE SHUTTLE ENGINE**

Figure 26 **APOLLO 8, FIRST MANNED FLIGHT TO THE MOON**

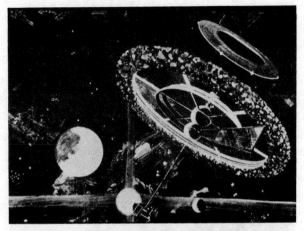

Figure 28 **EARLY SPACE STATION DESIGN**

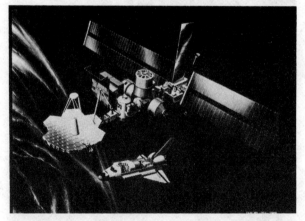

Figure 29 **CURRENT NASA SPACE STATION DESIGN**

Figure 30
**SIMULATED
SPACE SHUTTLE
FLIGHT**

Figure 31 **PANORAMA OF ROCKET HISTORY**

ARNOLD P. RITTER

A Leader in Space Technology

"Space": a simple word; one syllable, five letters. At the turn of the century, it still represented a mystery. It has been the object of studies by astronomers and philosophers for more than 20 centuries. Today, it has opened its doors, for mankind to travel into it and through it. It has no limits except those which man sets himself, as the late Dr. Krafft Ehricke stated. This was his guideline for his ideas and for his thorough studies to determine the solutions which would make his ideas come true. These studies were not out of fantasy novels, but were totally scientifically founded, as was the consultation by Professor Oberth for the film *The Woman in the Moon* in the '20s.

Space is not a simple word anymore, but encompasses all branches of science and engineering, and also philosophy and art. Krafft worked his way through every one of these fields, and one was always fascinated, in conversations with him, with the thoroughness and strength upon which his reasoning and conclusions were built.

I met Krafft Ehricke in 1954 at the General Dynamics/Convair Division in San Diego. Our families became close friends. Later, when Krafft was program director of the Centaur upper-stage space system, I became a member of his staff. The Centaur represents a particular milestone in Krafft's life, and at this point, I would like to say a few words about Krafft's path to his Centaur system and his follow-up activities.

Krafft devoted his interest to astronomy and space already in the early years of his life. At the age of 12, he went to a movie theater to see the film *The Woman in the Moon*. This film caused his ultimate decision to make astronomy and space his profession for life. During his college years, he was drafted. However, his brilliant mind in dealing with rocketry and space travel was soon recognized, and he was considered to be more useful to work along those lines. He was transferred from the Eastern Front, after about two years of service on the Western and Eastern fronts, to Peenemünde.

Peenemünde had been established in 1936 and was under the military direction of Gen. Dr. Walter Dornberger and under the technical direction of Dr. Wernher von Braun. Dr. Dornberger met Krafft in Berlin and brought him directly to Peenemünde. Their minds met and they became friends. Krafft asked to be assigned to some hardware job. He did not wish to deal solely with analytical studies. He was assigned to the propulsion system and saw in his activity a great chance for gaining detailed knowledge about a real missile, including design and manufacturing. Toward the end of the war, von Braun and Dr. Dornberger were able to separate themselves, with a team of about 135 key people from Peenemünde, and place themselves in the hands of the Americans. They were accepted and sent to El Paso with several V-2 rockets for demonstration and launch. Krafft belonged to the team, but had to make a tour through Berlin to bring his wife and daughter to safety. He was sent to El Paso after a corresponding delay. Specialists, of whom the von Braun team was a part, were taken to the United Stated under "Operation Paperclip," with a five-year contract. After the fulfillment of the term of the contract, von Braun was assigned with his team to the Army at Redstone Arsenal, Alabama, for the development of an intermediate-range missile.

Krafft did not want to continue on ballistic-missile development. He was looking for a chance to become involved in space activities. He joined the Bell Aircraft Company in Buffalo, after three years in Huntsville, which had indicated plans to enter into space programs. These plans did not materialize; however, Krafft learned about the Atlas development at the Astronautics Division of General Dynamics. He realized in this system a potential first stage for a two-stage launch system, which could place a payload into an Earth orbit or into an interplanetary trajectory. Herein, he saw his chance to go into space, and therefore he initiated studies of potential upper stages with the limited data of the Atlas available to him. The studies were supposed to provide him a sound basis for a discussion with Dempsey, the president of the Astronautics Division. Krafft hoped that the meeting with Dempsey would come about, since it was vital for him for the fulfillment of his dream, of the first step into space. It did; Dempsey was impressed, and Krafft became a member of a key team at Astronautics.

I would like to quote at this point a few sentences by William H. Patterson, former president of Astronautics: "Thanks to one of the most creative technologists in the nation at this time, Krafft Ehricke, we had a base to start from. Krafft had done in-depth analysis on a variety of upper stages. They included many fuel combinations including hydrogen-fluorine, gasoline-oxygen, ammonia-oxygen, hydrogen-oxygens, and others. Krafft had optimized each mission using the most efficient propellant combination with each." The choice was hydrogen-oxygen. For obtaining flexibility in placing payloads into respective final or transfer orbits, it was necessary to require multiple engine start capability. The anticipated system therefore entered into at least two new fields of technology: liquid hydrogen and liquid behavior in zero gravity.

The Father of the Centaur

After extensive studies, the first Centaur concept was completed in October 1957 and a proposal was submitted to the Air Force in December of the same year. In addition, Krafft gave a verbal presentation to the ARPA (Advanced Research Project Agency), newly established by the Air Force. This was again a typical Krafft presentation. It was objective, convincing, founded on thorough scientific and engineering studies, pointing to foreseeable problems, but showing that they would not be unsurmountable. He gained the confidence and respect of the audience. He was made aware that Pratt and Whitney had worked under classified contracts on liquid hydrogen and had developed a pump for feeding hydrogen into the combustion chamber of a rocket engine. ARPA recommended use of this system for Centaur. The Centaur propulsion system was modified correspondingly, and a second proposal was submitted to ARPA in August 1958.

With this proposal, Krafft won a contract for Astronautics. The contract contained the typical constraint: Develop the Centaur within the framework of the state of the art. This limitation appears to contradict the fact that new technologies had to be explored. Under the stated constraint, the initial funding was limited.

Krafft submitted, with a certain degree of humor, and got to work. Krafft was named the father of the Centaur. This gave him the satisfaction of the recognition of his deep connection with the Centaur, because with the Atlas/Centaur system, he would have made the first step toward space exploration possible. Krafft said: "The Centaur program is the bridge through which we make the transition from missile to space technology." The anticipated missions included the placement of a satellite into a synchronous Earth orbit, and onto the lunar surface. Also investigations of Mars and Venus with Atlas/Centaur-launched satellites were considered.

The Centaur development was initiated as an R&D program. When the system would have to be operational was left open. The only worry was, what incident during the initial phase could result in cancellation of the program?

The first flight was on May 8, 1962. A weather shield broke off, causing damage to the tank and an explosion after about 54 seconds of flight time. This incident, in addition to some other criticism, necessitated an extensive review and defense of the program. Krafft made a statement before the Subcommittee No. 3 (Space Sciences) of the Committee on Science and Astronautics of the House of Representatives (May 18, 1962). Krafft reviewed the total program, beginning with the first prototype prior to the contact with Pratt and Whitney, and followed with the second prototype, which resulted in the contract.

I would like to quote parts of his statement, to characterize Krafft's view of the Centaur program as a significant learning phase for future, larger systems. The first paragraph is an overview of the subjects he would cover in his statement:

In my prepared statement, I will review areas of systems concepts, mission objectives, engineering research and development, management, schedule, and cost of the Centaur program during the period of November 1958 through January 1962, during which I served as program director. I will also briefly cover the preceding period, from early 1956 on, the period of conception of Centaur. The Centaur concept was perhaps the result of the most thorough systems and mission analysis of any space project at that time. I was responsible for both the direction of the conceptual work, and for the GD/A program direction during the first three years of the Centaur development.

Krafft then outlined the study objectives, of which I would like to quote the following:

The multi-mission capability I spoke of, was studied with the objective to define and provide a first-generation capability for three of the most important mission classes at the beginning of an astronautics technology:

High-altitude satellites in the 8-hr., 12-hr., and 24-hr. orbits for the purpose of global surveillance, early warning, and global communication;

Launching of instrumented space probes to the lunar surface and into the inner solar system, primarily to Venus and Mars; but solar probes inside the orbit of Mercury were also considered, using the solid third stage, the same as we postulated for the lunar landing mission, since the energy requirement for both missions is the same;

Establishment of a small manned orbital laboratory for a crew of three to inaugurate systematic preparations for deep space missions of space ships.

These objectives show the orientation of Krafft's mind toward a logical, sequential exploration of space, including the concept of initiating a deep space mission for a space station, rather than from the surface of the Earth.

Following the summary of all study objectives, Krafft discussed the topics quoted in the first paragraph of his statement in great detail. I want to quote his comment on the concept of "state of the art," under which the program started:

> All this, and other necessities were to be added later as more funds became available. This is somewhat characteristic of a state of the art program approach. The purpose here is simply to "trigger" new lines of growth, get babies born who later on develop into full-fledged personalities. In such cases, it is difficult to predict precisely what problems one will find, what delay they might cause and what additional funds they will require.

In the concluding paragraph, Krafft gave credit to the GD/A team, the associate contractor, subcontractors, and NASA's management. I would like to quote the following sentence from this paragraph:

> In particular, I am, through my personal association, very much aware of the unexcelled capabilities of those among my former colleagues at MSFC [Marshall Space Flight Center] who are associated with the Centaur program under the direction of Dr. Hueter and Mr. Evans.

The presentation was a factual and honest review of the program, which detailed explanations of problems and approaches to solutions. It won the confidence of the audience and secured the continuation of the program. The management was transferred to the NASA Lewis Research Center. A new step-by-step development approach was formulated. The Centaur grew to a very reliable high-energy upper stage system, and together with the Atlas, carried GD/A through the extremely meager years of the late '60s and early '70s. GD/A survived these years thanks to two fathers: Charlie Bossart, the father of the Atlas, and Krafft Ehricke, the father of the Centaur. The multiple start capability of the Centaur was most significant for the mission of the second Helios, which required seven Centaur engine starts in flight for acquiring the planned heliocentric orbit. The Helios is a German-developed satellite for solar research.

The Vision of Krafft Ehricke

If we take the popular definition of "genius," which says that genius is 90 percent diligence, we can say that it applies to Krafft to that extent, that he was a person who worked more hours in every day than anyone else. Jim Dempsey was worried about his health, and said: "I want that man to stay alive for a long while, but I just cannot find a way to make him slow down. He will work at his desk right through the night." He once worked through more than one night, as an experiment on himself. He had taken pills which would keep him awake, to determine how long he could work productively without rest. He made it through two nights and three days in his office, and was satisfied with the sandwiches his secretary brought him. He completed the investigation which he had wanted to accomplish and then willingly took a rest with the satisfaction of having done so.

Krafft carried deep within himself an infinite source of ideas and the never-ending determination and devotion to work the ideas out in detail. He never considered an item as an item per se. He always saw an item as a part of an entity, which was ultimately space and space exploration. This is reflected in his vision of Centaur as the bridge from ballistic missiles to space technology, as mentioned earlier.

It was characteristic of Krafft's personality, that he could and always did extrapolate the present into the future, or, in other words, he saw beyond the present. But he never looked into the future from the standpoint of science fiction. In opposition to this, he had the energy, drive, endurance, capability for system coordination, and every other discipline required for learning and expanding the knowledge and depth of insight necessary for explaining the present and for talking realistically about the future.

Krafft's learning consisted of self-teaching, in the sense of translating everything he read into his way of thinking, into his way of seeing cause and effect, of seeing phenomena. Through this process of learning, he became a fascinating speaker and lecturer. He also learned through conversations with experts on a give-and-take basis. Krafft could always give as much to his partner as his partner could give to him. All conversations with Krafft were rich and animating. Krafft's memory was extraordinary, and he always had numbers and facts on hand for explanations and for carrying out numerical estimates to substantiate his statements or those of others. Krafft was not persuasive, he was *convincing*. All of Krafft's works, verbal or written, were productive and creative. In particular they were positive, in contrast to the widespread negativism of any news media in the free world. Krafft was unpolitical. He was a strong leader in space technology, by showing the way for a unified program for mankind within the framework of freedom, without the request to be called a leader. In contrast, there are too many who want to be a leader for the sake of being a leader, and who do more harm to the cause of the survival of freedom through their propagandistic action than could be tolerated.

I admired Krafft for the way in which he could speak about the near future or the future progress of science. He could do this only if he was absolutely honest with himself. He must have debated within himself whatever he wanted to say to the utmost degree, so that he could assert it as true, regardless of whether it were accepted by others.

I would like to connect the above considerations with a quotation from Krafft. Most of you probably know the saying: "My home is my castle," as a place of privacy. Krafft had modified this phrase and brought a deeper sense to it, which may touch upon his subconscious and indicate the key to the source of his self-confidence and his courage to express the results of his studies for the future of mankind. "To me, my mind is my castle. A part of it no one may enter—it is completely and absolutely taboo. *I must have this ultimate internal refuge, in which I am myself completely; only then I am really an individual.*"

In Search of the 'Seventh Continent'

The transfer of the management of the Centaur program to the NASA Lewis Research Center brought Krafft to a crossroad requiring a far-reaching decision. NASA requested, for a more effective continuation of the Centaur development, a project-oriented, self-contained organization. Dempsey wanted Krafft to become the director with total functional and engineering management responsibility. Krafft wanted to retain a staff and a separate study group for a continuation of his studies for the advanced systems functions of NASA, primarily MSFC and NASA headquarters.

Up to that time, Krafft had always responded to requests for proposals and always won study contracts for advanced systems. These studies were a world in themselves for Krafft, since his mind was oriented toward the future of space exploration. He conducted these studies with a thoroughness which was not equaled by anyone else, according to the testimony of NASA officials. Krafft reached the depth in his studies, among other factors, through attention to secondary or side effects.

Let me give you an example which is somewhat on the humorous side. In a reply to my good wishes for his birthday, he quoted the time and distance traveled during the preceding year. He started out with the picture everyone normally has in his mind, namely that the Earth moves in a circle around the Sun, and since we know the radius and time, we can express our year's trip in any dimension we like. But now came the correction: The true motion of the Earth is not circular, since in galactic space, the Sun, and with it the planets, move around a mass center and therefore the Earth moves along a spiral. He applied the correction to the distance he had traveled in the year, and thereby made his thank-you for my birthday wishes more complicated for me. I was grateful for the lesson, however.

NASA had a different reaction to the degree of depth in Krafft's studies. To his surprise, he found in one of the specifications for the research task a page limitation: The study to be submitted shall not exceed 60 pages. Krafft took note and conducted the study in his usual manner. The result was 90 pages. With the help of the photographic department, the size of the type was reduced so that the number of the pages remained below 60. Of course, Krafft knew what the page limit meant. NASA took his interpretation humorously.

The preceding outline was intended to illustrate the deep devotion of Krafft to the advancement of space technology. On the other hand, I can now, from my experience during the last two years, fully understand Dempsey's attitude, that full management responsibility cannot be combined with the degree of scientific studies as conducted by Krafft. The combination which Krafft wanted was unfeasible from Dempsey's viewpoint and giving up the advanced studies was unthinkable from Krafft's standpoint. Krafft was convinced that the reaction to his statement before the Subcommittee for Space Sciences, which demonstrated a new and reasonable approach, meant a secure future for the Centaur and that someone else could carry on management responsibility. He decided to pursue his advanced studies, but remained available as a consultant.

Ultimately, Krafft realized that he had to separate himself from the industry to establish his own company and to become a consultant, lecturer, and, first and foremost, a research scientist for the advancement of the space technology.

It is interesting to note that the film *The Woman in the Moon* had caused his final decision to devote his life to space sciences, and that his last major work was the detailed and in-depth outline of lunar colonization. Krafft had expressed the view earlier, that space exploration is of common interest to all mankind and should lead to peaceful cooperation on Earth for finding ways to reach other planets. The colonization of the Moon—the "seventh continent" as it was named by Krafft—shall be a joint effort, without national boundaries on the Moon. His book demonstrates Krafft's concept of the "Extraterrestrial Imperative": Mankind is destined for peaceful extension into the Universe.

Convair Astronautics

Krafft Ehricke played a vital role in developing the giant ICBM Atlas. he is shown here in 1957, holding a model of the Atlas.

Adult figures played a vital role in developing their own identities. Abraham Lincoln was a haunting and haunting model of the scene.

Lessons of Peenemünde and Other Historical Crash Programs for Today's Urgent Science Mobilization

Afternoon Panel
June 15, 1985

UWE PARPART-HENKE

The Question of Scientific Method

I want to start out by simply recounting one element of our association in the Fusion Energy Foundation with Krafft Ehricke. It did not come about directly as a result of his work in space-related matters, but on a rather broader subject. I believe my recollecton is correct that we first got in touch with Dr. Ehricke when an article appeared in the German daily newspaper *Die Welt,* in which he launched a pointed and direct frontal attack against the "Limits to Growth" philosophy that was being expounded by the Green Party in Germany and by similar kinds of organizations around the world, going back to the 1971 *Limits to Growth* book published by Forrester and Meadows at MIT. This philosophy was so contrary to Krafft Ehricke's entire outlook that he felt it was absolutely necessary to say in print and in very forceful ways why and how he disagreed with that way of looking at the world.

137

In light of Mr. LaRouche's remarks this morning about what defines a successful program, what is the conceptual depth and the conceptual breadth of a program such as the Strategic Defense Initative and other programs that we are now contemplating, it is absolutely critical to realize that it was, ultimately, Krafft Ehricke's broad philosophical outlook that there are no limits to growth, that any kind of thinking of that sort will necessarily lead us in the wrong direction, that basically defined his approach to the specific technical problems that he tackled as well.

What I want to do in my presentation is to contrast two types of approach philosphically, epistemologically, that is to say, from the standpoint of the theory of knowledge, to the kind of thinking that ultimately finds its way into large programs, like the Manhattan Project, the Apollo Project, or the Strategic Defense Initiative program. I want to point out in particular that one cannot simply see these as technical organizational problems or technological problems, but one has to get some understanding, of what is the broadest cultural background that defines the possibility of the successful development and execution of such large programs.

The Prandtl Film

The film that I want to show you now was made mostly in the 1920s and issued in 1927. Its title is *Generation of Vortices in Water Flows*. Such films were used for teaching students at the universities the characteristic features of fluid dynamics. This film was put together under the direction of Ludwig

Prandtl, director of the Kaiser Wilhelm Institute for Fluid Flow at Göttingen, who is probably the single most significant researcher in this century in hydrodynamics and aerodynamics research. It is the Prandtl approach to these problems of fluid mechanics and fluid dynamics, which I want to use to exemplify for you the type of outlook and the kind of philosophy that has to find its way into the development of these large-scale research programs, if they are ultimately going to succeed.

Let's first take a look at these filmclips without much commentary and then go into the background.

Dr. Tietjen, who is mentioned here, was the co-author with Prandtl of what was probably the most influental book on fluid mechanics. The surface of the water has been sprayed with some aluminum, in order to make it photographable, and this is water streaming around a cylinder. You can see the actual vortex formation, which becomes large-scale after a short period of time. Some of you may recall the pictures taken on Jupiter by the space probe, which showed a very similar kind of phenomenon, the large red spot on Jupiter (Figure 1).

This is a closeup of how the so-called boundary layer rips off and develops the vortex, the fluid vortex. Then you see the back stream, the backflow around this cylinder. You can think of this as an airfoil, as a wing, and you can see the backflow creating the vortex. If that occurs on a large scale and in the center of an airplane wing, it leads to stall. I want to call your attention to the very critical role of the so-called boundary layer, very close to the surface.

Now we are looking at an elliptic cylinder in the so-called subcritical regime. The boundary layer is that sort of light, surrounding mass around the dark cylinder. Subcritical in this case means that the fluid flow is relatively slow and does not lead to rapid vortex formation. Now you see hypercritical flow around an elliptical cylinder.

This is now a closeup of the same situation, of the critical region. You can see how the backflow comes up and rips off part of the boundary layer.

Now the stream lines around the sharp object, like the tip of a knife. Because of the large difference in velocity on the left side and the right side, the vortex develops immediately.

Now we are looking at an airplane wing, an airfoil. The most interesting moment is the first moment, when you see the onset at the end of the first vortex, which then begins to rip off the boundary layer. You can think of the airplane starting out, and that's what you will get in the air. If you have ever been in an airport and are close to a 747 taking off, you know that these vortices hit you quite hard, and in fact smaller planes cannot take off in the wake of a large jet. In this case the wing is accelerated and then stopped again and only the flow is followed to see what happens.

Now a rotating cylinder is investigated, and because of the rotation the boundary layer is not ripped off early and no vortices are formed. You could build a rotating airplane wing, that might be fine, exept that does not seem very practical. You can see how the rotation prevents the boundary layer from being disturbed by the possibility of vortex formation. When it is stopped, the vortex forms immediately. Initially, the cylinder was not rotating, the vortex formed, and, when it started rotating, it got rid of the vortex formation. You can see, when it stops, that the boundary layer is ripped off.

By applying suction, you prevent the vortex formation. When the suction is reduced, immediately the vortex forms and the boundary layer is ripped off.

The Göttingen Tradition

The kind of research that led to this photography and teaching films started at the University of Göttingen around the turn of this century. Prandtl came to Göttingen in 1904 and initiated this kind of research, building the first sizable wind tunnel and similar apparatus, which made observations of this kind possible.

What I want to do, is review for you some of the broadest philosophical background to the kind of thinking, that enabled the researchers at Göttingen, at Berlin, and at Aachen in particular, to make the kind of breakthroughs in fluid dynamics and in aerodynamics, in the early part of this century, that made manned flight, ultimately supersonic flight, and then rocket flight a reality. I want to counterpose that to a different kind of philosophical tradition, which, if it had prevailed over the tradition that led to the work of Prandtl, would have left us in a situation where most of the developments that we have seen in this century, and especially after World War II, would have been either very far delayed or might not have occurred at all. What you see here, is a kind of derivation of the tradition with some specific emphasis on the geometrical type of thinking, that was characteristic of the Prandtl school and of the individuals, whose earlier scientific ideas led up to that (see table).

In particular, I want to make a few detailed remarks about perhaps the most influental and least known mind in this line of succession, Jakob Steiner. At the beginning is Gaspard Monge. Monge was one of the principal researchers at the French Ecole Polytechnique, at the end of the eighteenth century, and he pioneered a method of looking at differential equations, equations which define different types of complicated physical processes essentially from a graphic or a geometrical point of view. These methods proved extremely successful in the early work of the Ecole Polytechnique and then led to a situation, in which many of the students not directly of the Ecole Polytechnique of Monge, but students conceptually of these ideas, perfected this and were able to make enormous progress in a very short period of time.

Jakob Steiner was born in 1796 and he came to the University of Berlin, which had then just been founded by his mentor, Wilhelm von Humboldt. He came to Berlin as somebody who did not have a job; he knew a great deal of geometry and was convinced of his ability to solve the most difficult geometrical problems, but he did not have the kind of formal education that would have allowed him to become a professor in Berlin at that time. He could not even become a teacher at the high school level: In order to do that, he would have to pass a so-called state examination, and he tried that in 1822 after he had just come to Berlin.

He had the bad fortune that one of his examiners

in the field of philosophy was Hegel. Those of you who attempted to read some of Hegel's writings will appreciate two things that Steiner did: First of all, before he was examined in philosopy, he wrote a note of protest, saying that he protested the idea that he should be examined in the kind of obscurantism that Hegel's philosophy represented. Hegel then, as you might imagine, retaliated in the examination itself and wrote a report. The quote we have is that Hegel said, "Jakob Steiner concerns himself only with entirely trivial reflections." These "entirely trivial reflections" define the conceptual basis in almost every respect of the type of work, which led to the film of the kind that I showed earlier of Prandtl and his collaborators.

Steiner's so-called triviality in the mathematical field was characterized by the fact that he abhorred algebra, and he was also tested in algebra. The two things he was tested in were Hegelian philosophy and algebra. He flunked both of these tests marvelously. The quote from the person who tested him in algebra, was that his knowledge of algebra does not appear to go beyond the solution of equations of the second degree and he does not even seem to be very familiar with that. Equations of the second degree, are something that, perhaps unfortunately, people are now being taught at a rather early age. But in any case, his genius in geometry was recognized, and perhaps we don't know the details, but perhaps first recognized by Wilhelm von Humboldt, who founded the University of Berlin and was the minister of culture of Prussia for a while.

He had his youngest son educated in private by

Steiner after Steiner had been denied official cer-
tification as a teacher. The first book that Steiner
wrote on geometry, which became the principal
textbook in geometry at the University of Berlin
later on and in many of the German universities
and high schools afterwards, was dedicated to von
Humboldt and von Humboldt's method of thinking. What Steiner always stressed in teaching his
students was that there is a very close relationship
between the kind of creative playfulness that we
apply in geometrical contructions and our ability
to develop entirely new concepts; whereas, on the
other hand, algebra puts the mind into the kind of
straitjacket that does not enable the student at a
later point to apply himself creatively to new types
of problems. I don't want to review in any further
detail the career of Steiner. I want to point out that
in 1834 he finally got his appointment at the University of Berlin, because it was recognized that he
was an obvious genius in his field. His efforts to
become a professor were supported by Crelle, by
Bessel, by Dirichlet, and by Jacobi, who were than
the greatest mathematicians in Europe.

The opinions of Hegel and of some other mathematicians, who initially examined him, were
thankfully ignored at that point and he was made
a professor. In 1847-48, he became a principal
teacher at the University of Berlin of Bernhard
Riemann, and it is the work of Riemann and Dirichlet in the nineteenth century that really laid the
foundations for the work in fluid dynamics and
aerodynamics that developed the possibilities of
manned flight and of rocket flight later on.

Especially from the standpoint of the possibility

of supersonic flight, a paper that Riemann wrote in 1859 on shock waves—the kind of waves that are formed in a compressible fluid, be it a gas or any other kind of compressible fluid—proved extremely influential. It was one of the most important things to consider, when supersonic flight was contemplated in the period of World War II and afterwards. Contrary to many of the critics of Riemann, it was precisely the fact that he discussed so-called isentropic compression shocks in his 1859 paper, which proved to be most important and influential in the theory of supersonic flight. Prandtl's training in Germany was very much in the tradition of Riemann, and in fact in some of his first papers, he quotes Riemann in detail.

He had a student, who perhaps many of you never heard the name of, Adolf Busemann, who worked in Germany during World War II, then came to the United States after World War II. His ideas were the essential ideas that made supersonic flight possible by October 1947, when the first Bell X-1 plane crashed the so-called sound barrier. (A lot of things could be said about this notion of sound barrier; there really is no such such thing, and it in fact implies all the wrong things; I make a point of that, because it implies precisely that wrong kind of thinking, which we should stay away from.)

The Opposing Tradition

We counterpose now the geometrical tradition, reaching from Monge through Busemann, to the tradition of the people, who if their ideas had prevailed, engineers or other inventors might have invented airplanes and done various kinds of things with them, but physicists and mathematicians would have been able to "prove" quite rigorously that manned flight or flight heavier than air was impossible.

One of the people on this list is Theodore von Karman, who in his very early career, just about one year before the actual first flight heavier than air by the Wright brothers, proved to his own satisfaction (not to the Wright brothers' satisfaction), that flight heavier than air is impossible. This was based on the theory of air resistance, of so-called drag, a resistance of any fluid against an object being moved through it: the so-called impact theory, or resistance or drag, due to Newton and later on developed in more detail by Lagrange. One could perhaps say, somewhat ahistorically and facetiously, but nonetheless correctly, that Newton was the first to prove that flight heavier than air or any kind of flight—in fact it is not even clear, how birds could fly under Newton's theory—was impossible.

Prandtl makes the point in his famous textbook (actually written by Tietjen on the basis of Prandtl's lectures) by saying, that if it were the case, that drag or resistance increases with the square of the velocity, then under those circumstances it is extremely difficult to see how flight of any kind is conceivable. The way Newton arrived at this, is on the basis of this so-called impact or collisional model; i.e., thinking of an airfoil or even a plate injected into an airstream, and simply computing the impact and the forces of impact of the molecules that impact on this particular airfoil, that impact on any kind of object put into the flow. This way of thinking and von Karman's calculations that led him to believe that flight heavier than air was impossible, were based on that kind of impact model. Essentially, he said, the molecular pressure would prevent takeoff. You shall see later on, how this kind of thinking was quite pervasive even at a point when von Karman later on became one of the celebrated people, who allegedly had a lot to do with the development of aerodynamics.

The point to be made here is that this collisional and essentially statistical model of computing physical events on the basis of certain averages, averaged over particles and groups of particles and molecules statistically, has proved one of the most important barriers to a satisfactory development of theory not only in the areas that we are discussing here—fluid dynamics, hydrodynamics, etc.—but also in the equally important areas of the field of quantum theory, of plasma physics, etc., which are essential to the possibility of thermonuclear fusion.

These collisional and statistical models do not work, and it is only and precisely to the extent that they were explicitly rejected by the first line of tradition that I pointed out to you, that the programs we have been discussing can be regarded as possible and developable.

The essential idea that Prandtl had in 1904, is that if one were to try to describe the possibility of flight, using the very difficult differential equations that govern the flow of so-called viscous fluids (fluids that have internal friction), the so-called Navier-Stokes equations, then one would be faced with an impossible problem. One could experimentally perhaps define and determine the possibility of flight, but one could never quantitatively explicitly calculate the actual conditions, that make flight possible. Prandtl, rather than looking at an airfoil subjected to a stream of air as an airfoil injected into a viscous fluid, which mathematically is impossible to handle, separated the problem in two, from the standpoint of the geometrical type of thinking, which characteristically introduces as an essential characteristic of the geometrical continuum the singularities in this continuum. He said, on the one hand, we can look at the flow far away from the airfoil, the so-called free flow, on the basis on the very simple potential equations according to Laplace. These are trivial and relatively easy to understand differential equations, which have an immediate geometrical interpretation in the context of so-called conformal mapping theory.

Prandtl said the only area in which we have to consider flow that has internal friction, is in the immediate vicinity of the airfoil itself, in the so-called boundary layer, and that is that little white layer that you saw around the objects in flight earlier. In this area, we can no longer ignore viscosity, we can no longer ignore the internal friction of the fluid, in particular not, because we know on the one hand, that directly at the surface of the airfoil the flow is zero; i.e., the air, or the water, or whatever it is, actually sticks at the surface. A very small distance away from this, it is clear that it has attained already the velocity, which is equal to the free flow velocity. What we must look at is this critical boundary layer or what Prandtl called a surface of discontinuity, in which, over an extremely thin layer—which can in fact be thought of as arbitrarily thin—a very, very large difference in velocity is attained.

If we take into account the theory of this boundary layer from the standpoint of thinking of it as a surface of discontinuity, under those circumstances we can simplify the Navier-Stokes equations quite significantly. We are therefore able to give a quantitative solution to the problems of drag, of lift, and all of the other aerodynamical problems that are critical to discuss the possibility of flight.

Without the kind of work that Prandtl did—first published in 1904, and discussed by him previous to his coming to Göttingen, when he was a teacher at the Technische Hochschule in Hannover—without these kinds of discussions of the boundary layer problems, it is generally acknowledged today that a quantitive discussion of the possibility of flight would not have been available.

One of Prandtl's most important colleagues was Runge, a mathematician who developed many of the mathematical methods for calculating the problems in aerodynamics that Prandtl raised.

The Role of Felix Klein

I would like to make a few remarks about the role of Felix Klein, the teacher of many of the students in the late nineteenth and early twentieth century in Germany in mathematics and in physics, who at the same time was one of the most accomplished organizers of the total scientific technological and industrial enterprise in Germany. Klein had earlier made a name for himself by developing some very interesting and significant work in elliptical function theory, and in the 1890s he came to Göttingen as a professor and made it his task to try to define a research program for the entirety of the technical and scientific disciplines at the University, in close collaboration with Willamowitz, who was the senior faculty member in the field of *Altphilologie*, ancient languages with specific emphasis on Greek. Klein and Willamowitz jointly defined an outlook on research and education, which I think is uniquely responsible, in terms of its philosophy, for the advances that were made in Germany in that period. At the same time, Klein in particular enlisted, and in a certain sense forced, German industry into supporting this kind of research, both by financially supporting the research institutions, that were being built at the German universities, and at the same time allocating up to 20 percent of the total profits of the company for research and development.

Klein founded the so-called Göttingen Association, the *Göttinger Vereinigung*, in 1898. This was the group of professors at Göttingen who collaborated with the principal people in German industry. The Göttingen Association mandated that any industrial company that wanted to get the top students from the disciplines of physics or mathematics, or the engineering sciences into their companies, could not get that unless they could demonstrate that more than 20 percent of their profits had, in fact, been allocated to research and development. They were otherwise not found worthy of being supplied with that kind of manpower.

Klein, because he had a very close working relationship with the Prussian minister of culture, Althoff, was able to quite rigorously control this situation, and was able to force those companies that did not want to comply into a situation where their competitiveness was, in fact, severely hampered.

Now, whether or not one wants to use that kind of model in the United States today is, I think, something you might want to debate and think about. But, in any case, the basic point here, I think, is very clear: that industry must make its contribution not only in the form of financial donations, but in terms of an actual, in-depth commitment to research and development, so as to be able to collaborate with the most advanced scientific institutions, so that there is not this tremendous and unnecessary gap between theoretical and applied research. And that was Klein's principal purpose.

He was able to enlist the heads of *all* of the large companies, from Krupp, to Siemens, to M.A.N. Any company of any size in Germany in the period before World War I became, at one time or another, a member of the Göttingen Association and collaborated in this program. It's this which made the developments possible which have been discussed and reviewed during this conference.

Von Karman and Other Villains

Now, let me review, in contrast to this, the type of approach that was taken by this second group of people I had indicated to you. Some of you who have worked in the airplane industry and the space program, etc., may not only be surprised, but perhaps offended, by the fact that I single out Theodore von Karman as one of the villains in this story, even though he admittedly made some significant contributions in certain areas. Von Karman, himself a Hungarian by birth, was a student of Prandtl at Göttingen. And Prandtl was instrumental in providing him with a professorship at the technical university in Aachen, in the westernmost part of Germany. In the initial years still directly under the influence of von Karman and of Prandtl, between 1908 and 1911, von Karman did quite excellent work there. In fact, much of the type of work on so-called vortex streets, vortex formations behind objects, and fluid flow, is due to the early work of von Karman. During World War I, he was drafted into the Hungarian Air Force, and he then returned to Aachen in 1920, to resume his post.

It is not quite clear what happens to one if one is drafted into the Hungarian Air Force, but whatever happened to von Karman was not very good. The actual scientific developments and the scientific initiatives that he took after his return to Germany, I think, are by and large, to be judged quite negatively.

In 1922, he organized a conference, along with others, at Innsbruck, Austria, in which he was the first to propose, directly in opposition to the geometrical approach of Prandtl, a statistical approach to the theory of turbulence. It was as a result of the disagreements that arose out of that—they did not really come very much to the surface or very much into the open, at least in these kind of disputes, scientists often tend to be polite, perhaps *too* polite, rather than bringing out these differences for everyone to see—but in any case, Prandtl quite strongly disagreed with this approach. It was directly contrary to his own way of thinking, and to his own insight into what had allowed him to succeed.

Prandtl blocked the appointment of von Karman to a professorship in Göttingen in the early 1920s. At that point, a different development occurred in the United States, and now we shift ground a little bit and look at the scene here.

After World War I, it had become quite obvious that airplanes and similar kinds of high technology devices had already had a very significant influence in World War I, and might, in fact, become decisive if a new war were to break out in the future. At that point, various organizations of industry, as well as military organizations in the United States, realized that the actual level of physical science and of engineering science in the United States was abysmal, and attempted in the relatively shortest possible period of time, to remedy that situation. One of the principal protagonists—and there should be no question that his motivation was right, though, I think, badly executed—was Robert Millikan, who, in 1923, won the Nobel Prize for physics for his experiments with electron theory.

Millikan, at that point, or at least slightly later, became the leading physicist and, in fact, the leading organizer of the research at the California Institute of Technology. He collaborated very closely with Daniel and Harry Guggenheim, for the purpose of making money available for the development of research institutes, and also for the possibility of attracting researchers, primarily from Europe, and with emphasis on Germany, in order to remedy the backwardness of the United States situation as it existed under those circumstances.

In one way or another, it became known to Millikan that von Karman was getting disenchanted

with his position in Germany, and by 1926, negotiations started between Cal Tech and von Karman.
Initially, von Karman acted as a consultant in the
construction of the wind tunnel at Cal Tech, and
then later, in 1930, actually permanently moved to
the United States.

There's some background to this, which I want
to mention because it's important. About Millikan
himself, he did some useful experimental work, but
his philosophical outlook on the scientific enterprise was essentially diametrically opposed to the
kind of outlook that I have ascribed to Prandtl and
others. His autobiography—mind you, this is not
a biography I'm quoting, but an autobiography, so
it reflects his own way of thinking—starts with recounting a little story from when he is four years
old. He is playing with his two-year-old brother,
under their porch, in the dirt, playing with dust.
He says, my younger brother picked up a bunch
of dust and told me, "Well, *eat* it. One can eat this."
And Millikan says, I didn't believe that, and I told
him it's not possible, but my younger brother, at
age two, did not want to believe me, so I told him,
"Well, why don't you eat it yourself?" And the two-
year-old picked up the dust and ate it, and then
ran, screaming, to his mother.

That, says Millikan, is how he became a physicist.
That is how he was first convinced of the value of
the experimental method. Well, as I said, if I wanted
to slander the man, I might have *invented* this story,
but in fact, it is the first paragraph in his autobiography; so therefore, presumably, he was deeply
impressed by this and somehow *believed* this kind
of nonsense. Well, that's not how you become a

physicist, or anything else. That's how you become a *fool*.

But, in any case, the later developments are not that important. I just want to read you, however, just a little of a list of whom, later on in his autobiography, Millikan regards as his scientific heroes. This reads, in a certain sense, like a list of the villains, but somewhat amended, that I showed you earlier. He regards as the greatest genius in the history of science, Maxwell. He then lists Kelvin, Rayleigh, Helmholtz, Boltzmann, and J.J. Thompson. Now, mind you, this is a man speaking in the 1940s. There is not a single mention here of people whom, I think, we rightfully should regard as the greatest scientific geniuses of the nineteenth and the twentieth centuries.

Well, I'll let you ponder that. The problem is, that the scientific enterprise in the United States, even at a point when, quite correctly, it was realized that it was backward, then came under the guidance of an individual who had done valuable experimental work, but whose entire outlook and way of looking at the scientific enterprise, was so slanted and so wrong, so badly misguided, that it is no real surprise that his programs, in fact, did not prove particularly successful.

Now, in terms of the Guggenheim-Millikan enterprise, they decided that they needed somebody. The word they used, was "finding a scientist of ability, bordering on genius." They wanted to find a scientist with the ability bordering on genius, give him some money, and let him develop aerodynamics in the United States. And the one they found was von Karman.

Why did they hit upon von Karman, rather than Prandtl? Well, here's the actual quote from a letter: Harry Guggenheim had gone to Germany at that time in order to look for such a genius. He had gone to Göttingen, he had seen Prandtl's work, and for whatever reason, Guggenheim was impressed, and said to Millikan, well, "let's get Prandtl."

Millikan responded, "Dear Mr. Guggenheim . . . With respect to the suggestion which you made as I left your house, that we try to get Prandtl over here for a short time, I have talked the matter over at length with Epstein and Bateman. Both of them think that in view of Prandtl's advanced age [mind you, he was five years older than von Karman] and his somewhat impractical personality, he would be far less useful to us than von Karman."

And then, later on, a little footnote is added, where he makes some remark about G.I. Taylor and Britain. In fact, they preferred G.I. Taylor as well. That may not mean much to many of you, but to some of us who know about G.I. Taylor's work, it means something. But in any case, it says: "The other thing that speaks for von Karman, by the way, is that he is Hungarian in nationality. We have between us reached the conclusion, partially because of von Karman's nationality, that he would be the better person than Prandtl."

One of the most famous quotes that I have of Millikan, is also in his autobiography; this was right after World War I and perhaps understandable in the heat of the argument in some respects; he said, what we can't have in the United States is the German barbarism reflected in World War I, and we can't have people associated with scientific work, in

Germany at that time—which was true for Prandtl who had a great deal to do with the development of airplanes. Then he said, "we Anglo-Saxons have overcome these tendencies toward barbarism. The British Empire, after ridding itself of some of its worst excesses, has become the veritable model of freedom and development in the world today."

So, this was the person who brought a genius to the United States.

The gist of what was the outcome of this, you could see at the end of World War II. From 1930 on, von Karman was effectively in charge of all of aerodynamic research in the United States. There was really nobody who could have challenged, in any way, negatively, or otherwise influenced, what he wanted to do.

In 1935 there took place in Rome the so-called "Volta Congress," on aerodynamics and fluid dynamics, in which certain presentations were made, the primary ones by Adolf Busemann, whom I already talked about, and the other one by General Crocco, who was one of the principal aerodynamics researchers in Rome. Von Karman went to that congress, after he had been in the United States for five years and had gotten more money for developing aeronautical research at Cal Tech than the entirety of European institutes taken together. He came back with the impression that the Europeans were far ahead. And he made a report to this effect, but couldn't figure out why. He said, we seem to be doing what we should be doing, but somehow, we don't seem to be succeeding.

In particular, he was quite rightfully impressed with the fact that after four years of trying at Cal

Tech, they had built a wind tunnel that was operating at several hundred kilometers per hour speeds, and something like 5,000 horsepower. When he went to Rome in 1935, he found a supersonic wind tunnel operating at twice the speed of sound, and with 20,000 horsepower. So he came back and was shocked and made the determination that all energies must be mustered to develop this work better in the United States. Nothing came of it.

In 1938, the question of jet propulsion was first investigated in the United States. There was some suggestion that jet propulsion should be a good way of driving airplanes. A committee was called together by the National Academy of Sciences, under the leadership of von Karman and Millikan, with the able assistance of Professor Marks of Harvard University. And they delivered their report on June 10, 1940. The report said, in essence, gas turbines are no good for flight because they're too heavy. Well, several months before that, the first model of the Messerschmitt 262, the actual German jet fighter of World War II, had already successfully flown and gone through much of the testing routine.

Von Karman delivered a report of the impossibility of jet propulsion for aircraft, at the time when such aircraft were already flying in Germany! He later on apologized and said that he just put his signature to this report, he didn't really read it. And then he said that when the report was issued in 1938, he was in Japan. He in fact was in Japan in 1938; however, the report was not delivered until 1940, so that doesn't make much sense, either.

The Army Air Force, in 1945, was quite shocked

when they saw what they had found in Germany. Several people were sent over in 1945, to Germany, to investigate what was going on. Von Karman was one of them. He and another researcher from Cal Tech went to Germany, and then for long hours they questioned Prandtl in detail about what he had been doing. Adolf Busemann was questioned in detail about his ideas on supersonic flight.

After von Karman came back, he was asked by the NACA, the National Advisory Committee on Aeronautics, as well as by the Air Force, to deliver a report. And he wrote a report which said, we weren't really very impressed with what we saw in Germany. In fact, in many cases, the German work was good, but it certainly was not spectacular. Many of the things that have been praised, we were ourselves thinking about.

The Air Force did not issue the report. One of the top people in the NACA, Hunsaker, wrote a letter to von Karman, saying that this seems to be a rather self-serving and nonsensical report, and you will make yourself a laughingstock of the world if you issue it. For your reference, said Hunsaker, I will list to you precisely those areas in which the Germans *were* ahead in 1945, and in which we did virtually nothing: Supersonic research, missile research, rocket research, jet propulsion, swept-wing design, and so on and so forth. And just listed those areas, primarily in the field of aerodynamics.

So, the report was not issued, but von Karman was promptly charged by the Air Force to write another one, outlining the next 50 years of aerodynamical research for the United States. I don't know if that was ever written, or maybe it's a classified document. I hope it's so deeply classified that nobody will ever see it.

The Question of Method

In any case, that brings us to the fairly obvious conclusion. There's no question that the financial and material means at the disposal of the German effort in aerodynamics and related fields during and before World War II were in no way superior. What was superior and was different, was the type of outlook and the basic method that I have stressed here.

Von Karman is associated with the statistical turbulence theory and with the idea of using the classical hydrodynamic theory, making certain linear adjustments in it, in order to get away from the nasty singularities that plague this kind of research. He's associated precisely with the outlook which, if it is adopted in principle, will not allow any significant advances in the physical sciences, and has never, in fact, been responsible for the development of such advances. That is the very simple fact that we have to face.

It has nothing to do with Germany versus the United States, or anything of that sort. It has something to do with *method*. These points of method were shared by the people of the Ecole Polytechnique in France, they were shared by the group around Riemann, they were shared by the great hydrodynamicists of Italy in the tradition of Riemann, and they were shared by all of those researchers whose names I already mentioned, most notably, Prandtl and Busemann in Germany at that time. It is not a point, as Millikan says, of nationality.

In the postwar period, there are a number of important things to look at. I will not look at the rocket programs because they have been reviewed here competently.

This is the so-called sound barrier (Figure 2). I object to the word "barrier" because it implies precisely that kind of collisional approach. It has nothing to do with barrier; there is no barrier, there is nothing there. There is just air, like anywhere else. The point is, that if you get near the speed of sound to about 0.7 Mach, then under those circumstances the drag coefficient on the airfoil increases very steeply, exponentially, until you in fact reach the speed of sound.

The reason for that is the fact that through the development of shock waves, which affect the airflow over the airfoil, a certain amount of the lift energy is converted into shock formation. That energy is taken away from the lift capability of the plane, and under those circumstances you experience various kinds of instabilities and difficulties with the plane itself, which have to be countered simply from the standpoint of understanding the problem—of making the kind of geometrical adjustments, in wing design, or anything else, that are necessary to do that.

One of the principal adjustments in wing design that can be made, was invented by Busemann, the so-called swept-wing design, the arrow design. You can see here (Figure 3) how the critical zone for the development of shock waves that influence flight and lift negatively is at the 0° angle; that is, if the wing is at right angles with the fuselage, you get the onset of the critical area at 0.7 Mach and then the drag coefficient declines afterwards.

If you have a 60° angle of the wings, then not even half the drag coefficient develops and you get it also at a much later point; namely beyond Mach 1. And if you have a 70° inclination with the fuselage of the wings, then you get to a point, where you get a very low, very late onset of the critical phase. Also, the amount of reduction in lift or the amount of increase in the drag coefficient is not very substantial. It is there, it will always be there, because shock waves form.

Shock waves are real, as was certainly determined by these methods of research in aerodynamics, that were carried out in the 1930s and 1940s in Germany primarily under Busemann's direction in Braunschweig. They are not what Rayleigh had critically said, when he criticized Riemann's 1859 paper. He said, shock waves do not exist; what exist, are singularities in the mathematical formulation of the wave equations, but we cannot assign any reality to such singularities. All it means, is that we have failed to come up with a solution.

As Riemann said, these things *are* real, and he said it 50 years before Rayleigh made that idiotic criticism. It was precisely because of that realization of the reality of the shock waves, that when supersonic flight was studied in supersonic wind tunnels in Göttingen and Braunschweig, and later on in Munich, Lake Kochel, etc., that these things were taken into account.

Here is an interesting example (Figure 4). This is the Douglas D-558, which was developed simultaneously with the Bell X-1 as a supersonic design in 1945 before the von Karman mission went to Germany and interviewed Busemann and others. That was their design (a): a straight wing sticking out, so you have the 0° angle situation of before, a tail end that sticks up, just as in the old designs of aircraft in the subsonic range.

Then von Karman and others came back to the United States in the summer of 1945, and after the summer of 1945 the D-558 looked like (b). It was all of a sudden a swept-wing model with a swept-tail configuration etc.

In fact, one of the interesting stories that I learned several years back when visiting a scientific conference in Moscow, was from a Russian researcher who showed me a picture of one of the models for a supersonic passenger jet type that the Russians had acquired when they moved into the eastern part of Germany. "What do you think that is? he said. I said, "Everybody knows, that's the Concorde." But it was not the Concorde, it was a model built by Busemann for a supersonic jet, to which the Concorde design is identical—done in the late 1930s.

There is no mystery of any kind involved here. It is a simple and straightforward story, it's a question of method, both of scientific method and of method of organization. It's a question of assembling the kind of scientific team, which is capable, on the basis of the right kind of methodological approach, to find the mode of organization most appropiate to its goals, and simultaneously, as was pointed out by one of the previous speakers, setting your goals never with regard to so-called state-of-the-art designs, but in fact, setting them as far beyond as you can possibly do.

To the extent that you do that, you will be capable of changing this so-called state of the art rather than being stuck with it. What we have to do in any program, whether it is a crash technological development program or a basic research program, is to set our sight on the kind of goals and tasks that are way beyond what we initially anticipate the most immediate goal of the program to be. If that is not done, then we will not confront ourselves with the type of challenge that in fact is necessary in order for the scientific enterprise to succeed.

The lesson to be learned, is that we do not need state-of-the-art programs; that is nonsense, and leads to precisely the wrong approach. The cheapest programs are not state-of-the-art assembly programs; the cheapest programs will always prove to be those crash programs that look as far ahead as possible in order to accomplish the immediate task. This may appear to be quite expensive in the long run, if you have to bring in basic research and technology and design and all of that together into a program, rather than just saying, let's do the state of the art, on the basis what we have on the shelf. The latter is going to be the most expensive and the least workable approach, and I am afraid, to a significant extent, when we are talking about the SDI today, it is precisely that kind of approach to the situation, that is most problematical.

Von Neumann's Cost-Benefit Nonsense

Concluding on that, I have to mention one other villain, whose name was mentioned on the list and who had something to do, not so much with the scientific side of these developments, but had a tremendous influence on this organizational side, John von Neumann, another Hungarian-born mathematian, who also studied at Göttingen and later came to the United States in the 1930s.

I have no time here, to review von Neumann's career, even any significant aspects of it, but you probably know that he is associated in the minds of most, not so much with his mathematics and physics, but rather with his ideas in economic theory. In particular he wrote a book along with Morgenstern called *The Theory of Games and Economic Behavior,* viewing economic development essentially as a kind of competitive game between players, much as players face each other in a poker game. In fact the first paper von Neumann wrote on economics so-called was *The Theory of Parlor Games,* in 1928.

The next thing he studied in order to be able to model economic development in the late 1920s, was poker, and he invented a simplified version of stud poker and abstracted from that his basic ideas of economic development. Don't underestimate the influence of this nonsense. What has come out of that is the Rand Corporation, the Air Force Systems Command, and every single bit of so-called cost-benefit analysis optimization nonsense that we are suffering from right now. It is one of the principal obstacles to getting defined and pushed through the kind of crash program for the SDI that is desirable.

The other thing that has come out of it, is the famous McNamara way of "winning" the Vietnam War. You remember what that was: it was the body count method—cost benefit analysis applied to military strategy and tactics. You all were treated to that, most of you, I am sure, every night on TV: You had a body count, so many Vietcongs, so many North Vietnamese killed, so many Americans killed, the ratio looks good.

They made detailed analyses of how many people exist in each age group in Vietnam, to see how many people were being eliminated per day, and then the question was, how many troops do we have to put in to win on the basis of cost-benefit analysis? How much do we get out of it, if we put so many soldiers, so many tanks, so many this and the other things in? From the standpoint of linear programming and optimization analysis, how do we win? You can't win that way.

The principal strategic problem in military terms and otherwise in politics is the principle of the flank. The principle of the flank defies by its very definition the idea of cost-benefit analysis, and this has precisely to do with the unexpected, to put a tremendous amount of cost into one area, where it is unexpected, in order to be able to then succeed as quickly as possible. The very opposite of the kind of thinking, so much associated with von Neumann and much of the Pentagon thinking today, is what is called for under these circumstances.

If we keep that in mind, and let that be reflected in our political approach to these questions, we may have a chance.

THE GÖTTINGEN TRADITION	**THE VILLAINOUS TRADITION**
Monge	Newton
▼	▼
Jakob Steiner	Lagrange
▼ (W. v. Humboldt)	▼
G.L. Dirichlet	Helmholtz, Kelvin
▼	Rayleigh
B. Riemann (1859)	▼
▼	v. Karman, v. Neumann
L. Prandtl	Millikan
▼ (F. Klein)	
A. Busemann	

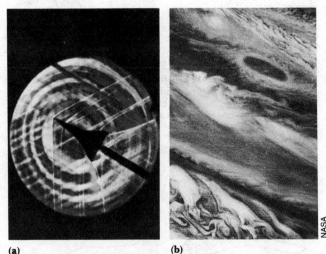

(a)   (b)

NASA

Figure 1 **HYDRODYNAMIC FLOW.** Study of the formation of vortices in fluid flow was pioneered by Ludwig Prandtl whose work influenced later advances in aerodynamics. These are photographs of air flow past a model of the Shuttle in a wind tunnel (a), vortical flow on Jupiter (b).

▼ Figure 2 **THE SO-CALLED SOUND BARRIER.** The so-called sound barrier has nothing to do with a barrier. If you get near the speed of sound to about 0.7 Mach, the drag coefficient on the airfoil increases very steeply, because shock waves develop that affect airflow over the airfoil.

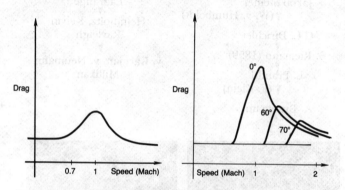

Figure 3 **SHOCK WAVES THAT INFLUENCE** ▲ **FLIGHT.** The critical zone for the development of shock waves that influence flight and lift negatively is at the 0° angle. If the wing of the plane is at right angles with the fuselage, you get the onset of the critical area at 0.7 Mach. But if you have a 60° angle of the wings, then not even half the drag coefficient develops, and if you have a 70° inclination you get to a point where you get a very low, very late onset of the critical phase.

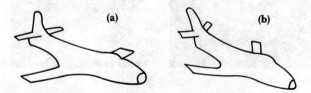

Figure 4 **THE D-558 DESIGN BEFORE AND AFTER BUSEMANN.** The Douglas D-558, which was developed simultaneously with the Bell X-1 as a supersonic design in 1945, had a conventional straight wing (a). After von Karman and others visited Germany and interviewed Adolf Busemann, their design was modified to his swept-wing design (b).

The Frontiers of Science Today: Plasma Physics, Beam Technology, and the Future of Space

Morning Panel
June 16, 1985

JONATHAN TENNENBAUM

The Coming Breakthroughs in Biophysics and Mathematics

It has been known for some time, since Edward Teller's October 1983 speech at the National Press Club in Washington, prior to President Reagan's announcement on March 23, 1983, of what has become the Strategic Defense Initiative, that the creation of such powerful defensive weapons as the x-ray laser depends on exploiting what Dr. Teller and others have called "new physical principles." It has also been stated that should certain details of presently highly classified experimental work in this area be released to the scientific community, every standard classroom textbook on quantum physics, especially quantum electrodynamics, would have to be thrown out of the window, or rewritten.

Dr. Teller and others have requested that, as much as possible, the new results be declassified, since exaggerated secrecy slows down the rate of scientific progress upon which depends the speedy creation of the kind of defensive systems necessary for our survival. After all, why should 99 percent of our scientific community continue to work with a form of physics which has already been proven, and known, to be wrong, by something less than 1 percent of physicists working in classified labs? Why should we not work with correct physics?

I cannot, and shall not, go into technical details of the very excellent work going on in this country on the perfection of the x-ray laser as a defensive weapon, as well as a scientific instrument par excellence. Nor will I go into the excellent work ongoing on gamma-ray lasers, and on the creation of condensed, laser-like forms of particle beams. What I shall do, is to indicate the nature of the new scientific revolution in the making, in which Dr. Teller's "new physical principles" play an important role. And I want to illustrate this revolution primarily with the help of examples from a field which, as some of you might be astonished to learn, is bound up most intimately with the physics of beam defense. That's the part of science known as "nonlinear spectroscopy," or otherwise as "optical biophysics" in the science of life.

This relationship between beam physics and biophysics and biology shows how idle and foolhardy it would be to attempt a crash program for the SDI on the narrow basis of developing high-powered beam technology per se. The only way to ensure the necessary rate of breakthroughs to get to operational systems in three years, and to get to a full-area defense of the United States and its allies within less than a decade, is an all-out, no-holds-barred assault on the fundamental frontiers of science as a whole.

The emergence of new physical principles at variance with textbook physics, is in fact long overdue. And it is no accident, that many of the central points of the new principles were already broadly identified by Mr. LaRouche and his collaborators and interlocutors around the Fusion Energy Foundation in the course of the late 1970s. Indications that Soviet scientists themselves were acutely aware, and sensitive to, certain of these problems, in particular certain implications of Bernhard Riemann's 1859 work on shock waves and of the crucial work on polarization of physical processes, provided decisive evidence that the Soviets were, and remain, engaged in intensive efforts to develop laser and particle beam technology for antimissile defense. Mr. LaRouche rightly pointed to this internal methodological evidence in terms of Soviet scientific method, as a much "harder" and more decisive proof that the Soviets were engaged in a crash program for beam weapon development, than the other kinds of evidence, such as, for example, submitted by Air Force intelligence's Gen. George Keegan around 1977.

Because in fact, when a nation decides to go for a crash program in science, a very crucial side of that is the jettisoning of certain kinds of ballast, of certain kinds of preposterous theories which simply don't work when you have to build things. In that sense, Soviet attention to Riemann's work, and its implications for a nonstatistical thermodynamics, is crucial evidence of the Soviet crash program. It would be just as if in the United States—probably the thing that would frighten the Soviets most—people would stop watching soap operas on television. It's quite obvious that a nation that's addicted to "Dallas" is not going to be able to defeat the Soviets. Therefore, I propose that's one part of our crash program. Turn off your TV. Start using your mind.

The Least-Action Principle

I want to sum up in a very condensed form what I think is one of the main new principles on the level of atomic and nuclear physics, which must be part of this scientific revolution. And that is, that we'll have to drop the notion, widely held in physics and science generally, that physical processes naturally tend toward what is known as "thermodynamic and statistical equilibrium." Most processes in the universe, and most emphatically those in biology, and also in high-energy-dense plasmas, can only be reduced or forced into statistical or thermodynamic equilibrium by such massive interference and intervention as we might justly say have "killed" the process.

Physical processes tend in general not toward equilibrium, but toward states best described as least-action, force-free states. This means, roughly speaking, that under boundary conditions providing for the undisturbed maintenance of the process, the process will pass through an ordered series of phase changes of increasing frequency and density. These changes will tend to be force-free, least-action in the sense that a minimum of action in the system is directed against the system. In other words, the system works for itself rather than against itself. What you would say in phase-space geometrical terms, is that the free energy transforming the process to a higher state always acts perpendicularly or orthogonally to the energy maintaining the process.

This principle of force-free least-action was already enunciated by Nicolaus of Cusa in his work on the isoperimetric theorem of geometry. Nicolaus of Cusa identified circular action as the primary self-evident geometrical manifestation of force-free least-action in physics. Leonardo da Vinci exploited this work of Cusa in his design of machines, and I would like to show you a few pictures of Leonardo da Vinci's drawings of machines illustrating this principle. What you want to do is build a machine which works for itself and not against itself. Man learned this very early, in inventing the wheel. It's better not to try to slide or drag things with friction, where the process is working against itself, but make a wheel which rolls and does not slide—a very simple principle, the same principle which later, around the French Ecole Polytechnique in 1793-1795, was the basis of Lazare Carnot's teaching of geometry in physics, and his definition of geometrical motions, which was his word for force-free least-action systems.

Figure 1(a) depicts circular action, identified by Nicolaus of Cusa as least-action. Geometrically, the isoperimetric theorem says that of all closed curves of a given perimeter, the circle is that which contains the greatest area, so corresponds to the most efficient possible action, whereas curve (b) works against itself to a certain extent; it is not least-action. Obviously, in geometrical terms, we have to build machines or processes that work like the first one, but not like the second one.

Leonardo, as you can see in Figure 2, never does crash programs on limited issues; as you see, at the left of the figure there, he's studying the chemistry of the human respiration process.

Figure 3 is interesting if you remember the film by Prandtl which Uwe Parpart-Henke showed yesterday on the generation of vortices in hydrodynamic flows. In what you see to the right there, Leonardo is studying exactly that question. FEF member Dino de Paoli has looked very carefully into all the drawings of Leonardo, and has seen that Leonardo actually draws all of the main forms of hydrodynamic flow and phenomena which were later studied by Prandtl. He was in a sense the father of modern hydrodynamics.

Crucial to Leonardo's study of least-action force-free systems, is his study of the way in which the human body works, and in particular the principles of least-action which are employed in the construction of the human body to obtain the maximum efficiency of work. In Figure 4, you see a comparison between the human hand and human arm and the arm of various other kinds of animals, monkeys, and comparing the way a monkey does work—as for example swinging from trees—and the change in the design and proportioning of the skeleton corresponding to different kinds of work which a human being does.

What we are doing now, is essentially mastering

the same principles of design of machines as Leonardo pointed out, as Carnot and Leibniz pointed out; however, we are doing it now, so to speak, on the level of atomic and nuclear physics, in the physics of plasmas. Indicative of these force-free least-action systems are present work on superconductors, present work on lasers, and present work on semiconductors, where you see exactly the kind of phenomena of maximum efficiency, zero or next to zero friction, next to zero loss, which in fact characterizes the efficient design of machines. Figure 5 shows in hydrodynamic terms what is called a force-free vortex, or Beltrami vortex. It's a vortex where every flow line is at the same time a vortex line, and it is around every flow line that the fluid is also actually rotating. That corresponds in plasma physics to an organization of current in such a way that the current does not to have to do any work against the magnetic field.

It's the same kind of geometry, as fusion scientist Winston Bostick, who is here, has pointed out, that you'll have in superconductors, a self-organization of a plasma so as to make a plasma, ionized gas, into a kind of superconductor. In Figure 6 you can see vortex filaments created in a plasma machine, which are force-free filaments and have the basic geometry which I showed before. These filaments carry enormous currents and focus energy to extreme intensities. The energy density in some of those filaments compares with that created by a hydrogen bomb explosion—in a device that can fit on a desk top! That shows what force-free least-action dynamics can do, since they are working for themselves and not against themselves, to focus energy.

Biological Systems

What I want to emphasize, as I said, is biology, because it's biology that gives us the most dramatic proof that in fact, in nature, in the universe, systems tend into force-free systems. The crucial thing, which is the crucial point about Riemann's 1859 paper, is that when we are talking about force-free systems, we do not mean something that just sits there. Our superconductor will conduct current, it seems, for a long time, but that's an exception. Biological systems are really the rule, or force-free systems which generate a whole series of singularities of increasing density. That's what happens when biological systems grow—that's what life is all about, it's growth.

If we look at photosynthesis, if you look at a leaf on a tree, it looks pretty peaceful, doesn't it—it doesn't make a lot of noise or anything, but if you were to calculate the rate of negentropic transformations going on in that leaf, when it has sunlight on it, you would arrive at billions of transformations per second in each of the minute photosynthetic organs—the chloroplasts—within each cell in the leaf. And yet, there is no noise, no vibration, there's little or no loss.

In fact, as I will take up in just a moment, studies done by Professor Popp at the University of Kaiserslautern in Germany on light emission from cells, have shown that plant cells, if you put them in the dark for a long time, emit a very weak light. It's extremely weak. The amazing thing about it is that if you damage the cell, the amount of light emission *increases*. In fact it increases when the cell is dying. How do you figure that out? The machine makes more noise when it's going awry, when it's not going right. In other words, the characteristic of biological processes is their absolute perfection in terms of efficiency. It's as if you were to walk into the middle of a huge industrial plant, a gigantic power plant of 2 billion megawatts, a steel mill and a machine shop, and you don't hear a single noise. That would be like what's going on in biological systems.

These are phenomena which are really the exact opposite of that which would be expected from the standpoint of statistical gas theory, which sees all phenomena as motion of molecules. In other words, the statistical gas theory approach, which Uwe Parpart-Henke outlined for you yesterday, basically looks at life as no more than jiggling—you are born, you jiggle a certain number of times, and you die. In other words, according to this view, life is just a "break dance" for molecules!

You see, the crucial epistemological point here is that you cannot reduce processes, and certainly not any interesting process, to fixed properties of fixed objects, and that is a tendency always of that school of thought known as reductionism, or Cartesianism—to try to deny that there is real change in the universe, and even more that it is change, and it is transformations, which are the whole substance of the universe.

Let me go into a couple of examples from optical biophysics. It is a very entertaining field, because we are challenged there directly with what we know is true in terms of shortcomings of present physics, but we are confronted with them in a very concrete form, because the life process comes out and says, "Hey, you'd better change your way of thinking!"

The first category of work which I want to mention is work on what is called "biophotons": very weak light emission from cells. There's a wealth of experimental work here. It goes back to the 1920s when evidence was accumulated that cells produce light, in particular ultraviolet light, of very specific characteristics and frequencies, for the communication between cells. In particular, there is the discovery of something called "mitogenetic radiation," which is just tuned ultraviolet light, which, according to experiments, allowed a light emission from one cell to trigger the division of another cell. Evidence accumulated that these light emissions played a crucial role in the coordination of cell activity in tissue. It was shown that this radiation in fact must be ultraviolet, because if you put a quartz window in between the two cells involved, it works, and if you put a glass window, it doesn't. Quartz lets ultraviolet through, and regular glass does not.

Studies have been made of this radiation. In the 1920s, one could not measure it directly—you did not have the photomultiplier tubes that we have now—but in the postwar period, this light emission has been measured directly. You put your cell in a dark room and let it quiet down; otherwise you have fluorescence, where a cell put out into the light will absorb that light and re-emit light, as you also have in nonliving processes. But if you keep the cell in a dark room, the fluorescence dies down, and you are left with a very, very weak emission of light. There is substantial evidence, that in fact this light, at least in large part, is generated by the DNA molecule in the cell, the double helix which we'll see in a moment.

How does this work? There was for a long time a paradox in genetics, I think it's called the "c-value paradox," which was one of the major paradoxes for the funny fellows who have been trying to prove that the living processes function like digital computers, with the DNA playing the role of the "computer program." One problem is that if you look at the sections of the DNA which can be traced toward particular genetic information, you find that's only a small percentage of the total base pairs, of the total molecules and atoms in the DNA. There are large sections, which have apparently a periodic structure, for which the geneticists could give no explanation. But since nature is force-free least-action, at least in undisturbed nature, if we don't hit it over the head, everything has a purpose.

The discovery of light emission from the DNA has solved this paradox to a large extent. The DNA is acting like a laser, that has a very particular geometrical structure, and properties that allow the molecule to absorb electromagnetic radiation of low quality in the cell, in the infrared region, and re-emit that radiation at a shorter wavelength, ultraviolet. That ultraviolet light represents a higher energy-density than the infrared light which was put into the DNA. This high-quality ultraviolet light, in turn, runs the metabolic processes of the cell!

Dr. Popp and his group have corroborated this remarkable hypothesis with a wide variety of evidence, including that provided by the structure of water in cells. What is structured water? Well, the standard textbook statement, "living organisms are mostly composed of water" is highly misleading, because the water in a living cell is not at all the same thing as the water in a drinking glass. It's the same old Cartesian problem again: We think too much in terms of objects instead of processes. In the inner fluid, the cytoplasm of the cell, the water is "polymerized" through the polarization of the water molecules, forming fibers which in turn create a network of channels which are able to conduct laser light with no loss. In other words, they are wave guides. And that water is structured in particular through the action of certain trace elements, especially minerals and metals in very low quantities in the cytoplasm of the cell. This vast network of "light pipes," it appears, conducts coherent light pulses from the DNA to various sites around the cell.

Now I want to mention one more line of research, with revolutionary implications. Many studies have been made on carcinogens, those substances which apparently cause cancer. If you try to understand carcinogens from the standpoint of ordinary chemistry, you'll go crazy, if you try to find a chemical property which these various molecules have in common and which makes them cancer-inducing. The group around Dr. Popp has done studies of the spectroscopy—that is, the spectrum of light emission and absorption by a broad range of molecules which induce cancer—and have found remarkable tendencies for similarities in the ultraviolet spectrum of these molecules. They may be shaped very differently, and they may have completely different chemical properties, but the ultraviolet spectrum is very similar. And we know that ultraviolet light plays a crucial role in all of the processes in the cell.

Electromagnetism:
The Case of the Moth

Other work, in particular work done by Philip Callaghan, who is an expert on insects and has written a book called *Tuning Into Nature*, has confirmed massively the extent to which we must get away from the so-called 'lock and key,' building-block notion of chemistry. When somebody asks, what do you consist of, you say molecules sticking together in various ways. Let me tell you, sticking together is not enough. You have to look at this thing in a completely different way.

What Callaghan did is observe very carefully the sense organs of insects, in particular moths. Figure 7 shows a certain approximation to a moth. You know that moths are able to sense each other through certain scents, certain smells, in particular sex attractants. Female moths will emit a certain kind of substance called a pheromone into the air. The male moth will pick up this scent, and will home in on the female moth. Philip Callaghan, who was a radar engineer during the war and designed radar antennae, was struck by the similarity between the way the male moth homes in on the scent, and the way in which a plane follows a navigation beam, going in kind of a spiral motion back and forth. He came up with the hypothesis that perhaps it's not really smell in the ordinary way, or that smell does not work in the way we think.

The usual theory of smell is the Cartesian "lock

and key" theory. You have a cell membrane, and
it has certain active sites, and you have the molecule
you are smelling, which has actually bound into one
of the sites chemically. That's how you are supposed
to smell different molecules, through chemical reactions.

Callaghan took electron micrographs and had a
look at the fine structure of the antennae on the
moth. This is a combined picture, you'll never find
all these different things right near each other on
the antenna. As an experienced radar engineer,
Callaghan found every single type of well-known
radar antennae reproduced on a smaller scale on
insect antennae: for example, pit antennae, log periodic antennae, phased arrays, etc. The difference
is that the insect antennae are much smaller—on
the order of about 10 microns. This would correspond to receiving signals in the infrared light range,
as opposed to the microwave range covered by radar antennae.

So the hypothesis was that what is going on here,
is that the moth is picking up an electromagnetic
wave, is picking up light. And these antennae are
actually tuned to specific frequencies. He took a
spectrograph, and watched the light spectrum from
these pheromones, these sex-scents and other scents
emitted by plants and so forth, and found very
particular spectral lines, particularly lines that are
strengthened very much when the molecules involved are vibrated in a certain way through an
acoustical signal, or are excited by ultraviolet light.

And then he was able to match these things, and
to do a critical experiment: Take the pheromone
from a female moth and put it in a bottle, close it

up real tight, and shine an ultraviolet lamp, or a flickering ordinary lamp, of fairly high power, onto that bottle; the pheromone molecules emit these lines, and the male moth responds as if there were the scent of a female moth right there. Yet there's no scent; there's only light. And if you turn off the excitation energy for the pheromone, the male moth doesn't see anything anymore—so you can turn the male moth on and off.

This also has dramatic implications, not the least for agriculture. What comes out of this, in fact, is the way in which our biosphere is organized, and the amazing specificity of biological processes—that one insect will fly only to one very particular species of plant, that one virus will only attack one particular species of cell, and so on and so forth—we may be unlocking the mystery of how this whole thing is organized! That means that if you want to control insects, why don't you just jam the frequencies that they are tuned to? And that could be a very revolutionary way of solving some of our problems in agriculture.

One last example is work on enzymes; enzymes are very crucial to these least-action, force-free processes. Because how is it that the chemical reactions, apparently, in the cell, occur at rates sometimes millions of times faster than we can get them to occur in the laboratory? It has something to do with these so-called enzymes; the geometry induced by these enzymes speeds up these processes and makes them force-free. Work by the French physicist Biscar and others has begun to unravel the "mystery" of protein enzyme structure, by showing that the various chains in the enzyme molecule function as tuned antennae, matched in such a way, as to focus electromagnetic energy onto active sites on the enzyme. It is found that certain frequencies of light greatly stimulate enzyme action.

Now, I want to get to the DNA, emphasizing the methodological implications of all this work in biology and biophysics. The point is, the universe does not consist of bumps and lumps; what the universe consists of is action, transformation. In fact, you can't even see anything without seeing a transformation—what you always see is a transformation. If there weren't any transformation, you wouldn't see anything. So you never *see* an object; you only *see* a transformation, a change.

The Challenge Facing Mathematics

What we must do is develop a mathematics which addresses that reality about the universe. In other words, we have to have an honest mathematics, a true mathematics. And that is very upsetting to certain scientists who want to be pragmatists, and say, "Look, my statistical theory works, I can calculate that and that, so what are you bothering me about? Why bother me with the question about whether my terms and my concepts actually correspond to something that's actually there? Don't bother me about that, because what I have, works!" Pragmatism.

To develop a true, a real, an honest, mathematical physics, with a greater approximation to the way the universe works, we have to realize, and take into account, the fact that the substance of the universe is process and transformation. There are no objects—there are no inert objects. In order to do that, we have to, of course, get rid of algebra, forget algebra. Once you understand something, you can do some calculations on it, but don't calculate before you understand it—a very simple principle. And start to think about how to develop a mathematics *of transformation*, and of *least action, force-free processes*.

Think of the circle—instead of thinking of it as an object, think of it as a method of representing to our eye the process of rotation, which is the simplest least-action process. Now, as many of you have read, if you think of—in a physical space—a simple, force-free process of rotation, and allow rotation to act upon rotation—rotate the rotation, so to speak—then geometrically that corresponds to rotating a circle. And if you act perpendicularly to the original, then that has the effect of folding the circle upon itself, and generates a diameter (see Figure 8). When you generate a diameter, what does that mean in terms of action? That means, by defining a diameter, I have now defined what half a rotation is; I have halved, or divided, the rotation. So in other words, you can start with simple least action (complete circular rotation), and by applying it again and again, you generate other forms of action.

And that's crucial—these very simple constructions with the circle, are really the basis for a revolutionary development in mathematical physics, which is not really new, was already begun by Nicolaus of Cusa and so forth, but it is now urgent that it be elaborated and used. That means, to return mathematical physics to the kind of *language for science*, that great languages, classical literary languages, were for literature. In particular, to go back to the principles which the great Sanskrit grammarian, Panini, applied to the Sanskrit language, in insisting, in the case of Sanskrit and all great languages, that the language is constructed of the verbs, not the nouns. In other words, nouns are determined by verbs, and not the other way. You don't determine a transformation by taking two objects and saying, "This goes to this," which is your modern mathematics, set-theory approach—an arrow between two objects. Rather, you define nouns as singularities of verbal transformation—generate them out of verbs. And if you see the system of roots in the Sanskrit language, it works exactly that way: Nouns are derived from verbs.

Now, the crucial thing is, if you start with the universe, and make the hypothesis—which is the only one coherent with the existence of the human mind—that there is a single universal principle of action in the universe, that gives you the action which corresponds to the transitive verb "to be," i.e., "to make itself become," in language: Now, what happens as the universe develops? Out of that simple verb, so to speak—other types of transformations are derived. The universe grows by multiplying and generating new types of verbs, new types of transformations. And the density of those transformations, increases as the universe grows. So the universe does not generate objects, it generates new forms of verbs, new forms of transformation. We require a geometry to describe this process—what is called "synthetic geometry," as opposed to both Euclidean and non-Euclidean varieties of axiomatic geometry.

The simplest way to do this is to take circular action; if you have a negentropic process, a least action process, which is generating singularities, how do you map that? The simplest way to map it is by a cone, where you imagine a circle which is growing, circular action which is growing. And growing means, greater density of singularities, greater density of new kinds of transformation (see Figure 9). Now, types of transformation occur in species. At a certain point in the process, a completely new, higher-order kind of transformation is generated, so we have to rather think of mapping the process as a series—as you see here, and as Mr. LaRouche mentioned for the case of the economy—a series of hyberbolic cones, in which each one of them is an increasing density of singularities, and between each one a new species is introduced (see Figure 10). So you get an apparent discontinuity, each time a new species is introduced.

That's the direction in which we have to go. And we have to map that thing on a sphere, because the universe is closed, and what you're doing is dividing a closed process—a unity—and elaborating it. So you have to think of it in terms of spheres.

The Golden Mean Proportion

I want to very briefly give you an approximation of how we can use synthetic geometry to grasp how physical processes work. And the principle is that of Leonardo. How is it that Leonardo was one of the greatest painters in all of history, and also one of the greatest scientists? What defines a scientist?

Rigorous thinking! And what derives from rigorous thinking? A very acute eye for the kinds of things that most people overlook, but are right in front of their nose. That's the eye of a great painter; and if you read Leonardo da Vinci's writings, he will explain to you, in terms of painting, that every painter must be a scientist, and every scientist a painter. Because what are you doing when you paint? You are creating the process yourself, re-creating the process which you are drawing. And when you create something, you understand how it works—very simple.

The first thing is, the role of the golden mean proportion and the pentagon, which characterize, as proportions in visual space, living processes as opposed to nonliving processes. Figure 11(a) shows a snowflake, which is a crystal—nonliving—a hexagon, six-sided figure, and Figure 11(b) shows a small creature from the sea, a microscopic creature, which has pentagonal symmetry. And already Kepler had pointed out that normally in inorganic nature, you never find this five-pointed symmetry; five-pointed symmetry characterizes life. Of course, somebody could build a pentagon or something—we could build a pentagon—but that's life playing tricks.

If we look at the DNA molecule, the double helix in Figure 12(a), the crucial thing for somebody who has a nose or an eye for geometry, is the fact that a DNA molecule in the "B" form, the active form of DNA, has a whole series of pairs, base pairs, indicated by those rungs—adenine, guanine, and so forth. Each single rotation, each cycle of the double helix, corresponds to exactly 10 base pairs, weapons; for example, how you get a laser beam which is so tuned that it can pass through the atmosphere with no loss—least action, can be absorbed in the target, and transform that target in the most efficient way.

I want to show you a couple of pictures on the broader kind of science mobilization that we are going to have to launch in the United States. I wanted to emphasize the theoretical side, because that's the motor of everything—it's the thinking which is the motor—but we also are going to build things.

First, there is the creation of life where there isn't any.

This is the desert (Figure 16); and our instinct should always be, when we see a desert, that we should plant things there. So what we should do is, do something like center-pivot irrigation agriculture, where you set up a system of irrigation, putting in the fertilizer and so forth, to transform desert areas into green areas.

There is only one problem: namely, this desert is on Mars. You'd just have to go 55 million miles to get to your new farm area. Watch out for con men who try to sell you land up there!

Of course, we also have a lot of work to do on the deserts on the good, old Earth.

Another one of the challenges that is part of our science mobilization is the creation of energy—namely, the mastery of thermonuclear fusion energy.

The Nova laser in Lawrence Livermore National Laboratory, for example, will focus its energy on a tiny pellet, raise it to temperatures of more than a hundred million degrees, and get fusion energy. We can have fusion energy by the early 1990s—that's part of the crash program—if we push this thing through.

Of course, we also have the challenge of the SDI, our crash program for defense. And the applications of laser technologies for revolutionizing industrial production here in our industry.

This is what we have to get back to, that sense of challenge. Look at the United States, look at the world, at your child, and ask: What's the right challenge? To have this child grow, and be well-behaved, and get rid of this kind of nonsense, this kind of decay.

So, to sum up: Launch a science mobilization, based on fundamental new scientific principles—that's what we have to do to mobilize this nation. And I think the key word, or model, of what this means, is Schiller's poem on hope, where he says, "Zu was Besserm sind wir geboren"—we were born for something better; we were born for something better than the kind of misery that we have on this Earth. And the answer is science.

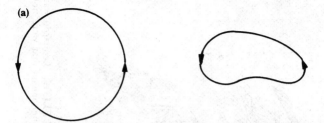

Figure 1 **CIRCULAR ACTION AND THE ISOPERI-METRIC THEOREM.** Of all closed curves of a given perimeter, the circle is that which contains the greatest area.

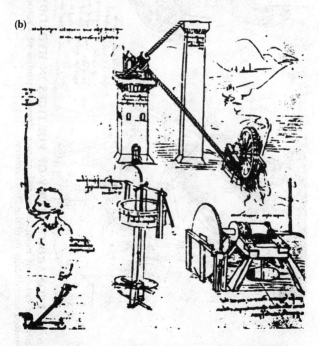

Figure 2 **LEONARDO'S STUDIES OF PUMPING DEVICES.** Leonardo's studies of pumping devices included human respiration.

Figure 3 LEONARDO'S STUDY OF VORTICES IN HYDRODYNAMIC FLOW. Leonardo was actually the father of modern hydrodynamics, drawing studies of hydrodynamic phenomena that were not recognized scientifically until the 20th century.

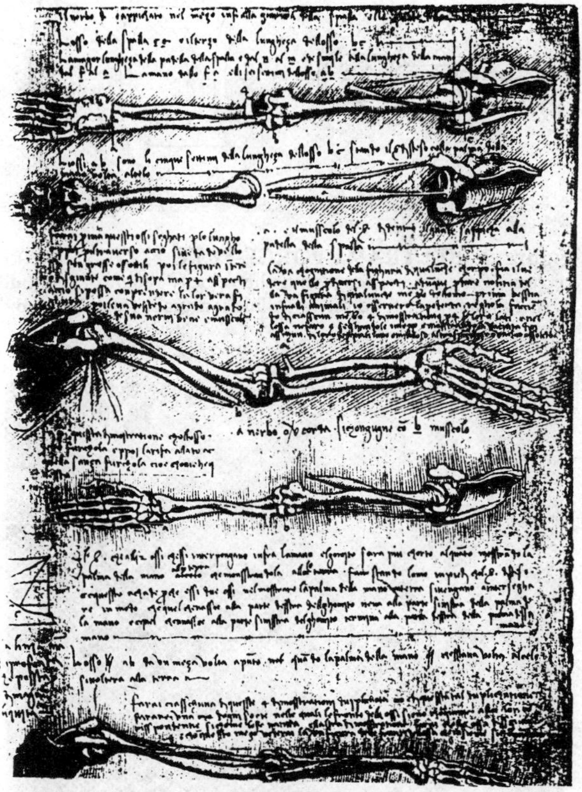

Figure 4 **LEAST ACTION PRINCIPLES IN THE HU-MAN BODY.** Leonardo compared least action principles in human versus animal limbs.

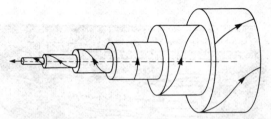

Source: Winston Bostick, "The Morphology of the Electron," *International Journal of Fusion Energy*, Jan. 1985, p. 38.

Figure 5 **BELTRAMI FORCE-FREE FLOW.** The direction of the arrows represents the velocity vector of the fluid as well as the vorticity vector. The fluid is actually rotating around every flow line.

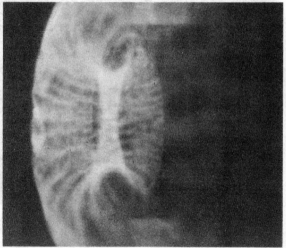

Courtesy of Winston Bostick

Figure 6 **VORTEX FILAMENTS CREATED IN A PLASMA FOCUS MACHINE.** This is a photograph of the tiny plasma filaments created in the plasma focus fusion device. The filaments carry enormous currents and focus energy to extreme intensities—an example of force-free, least-action dynamics.

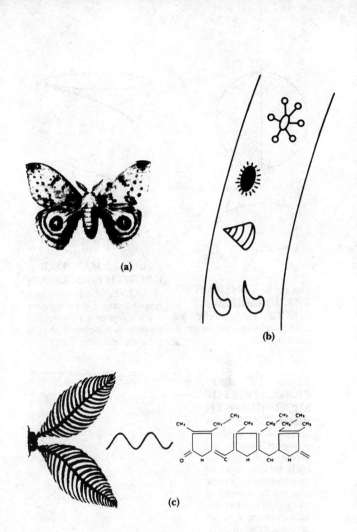

Figure 7 **ELECTROMAGNETIC PROPERTIES OF MOTH ANTENNAE.** A schematic of a moth is shown in (a), with various moth antennae designs in (b). Philip Callaghan has found that moth antennae are electromagnetically tuned to specific frequencies emitted by particular plants (c) as well as particular insect scents.

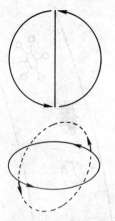

Figure 8 **GENERATING THE DIAMETER OF A CIRCLE.** Folding a circle on itself generates a diameter, a straight line.

Figure 9 **MAPPING A GROWTH PROCESS ON A CONE.** Circular action mapped onto a growing cone is the simplest way to map a least-action process that is growing, creating more singularities.

Figure 10 **MAPPING A SERIES OF SUCH GROWTH PROCESSES.** A process like a growing economy, where the density of singularities leads to new types of transformations, can be depicted by a series of cones, each of which has an increasing number of singularities, leading to new types of transformations.

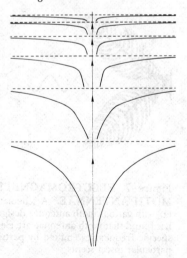

(a)

(b)

Source: *Fusion*, May-June 1984, p. 26

Figure 11 **PENTAGONAL AND HEXAGONAL SYM-METRY.** Living processes are characterized by pentagonal symmetry, while inorganic matter has hexagonal symmetry.

(a)

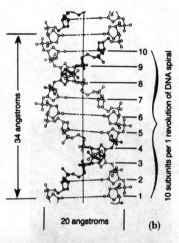

(b)

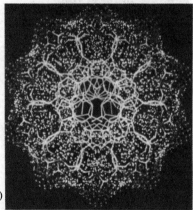

(c)

Figure 12 **THE GEOMETRY OF DNA.** The DNA double-helix is shown in (a). One 360-degree turn of DNA measures 34 angstroms in the direction of the axis (b). The width of the molecule is 20 angstroms, to the nearest angstrom. These lengths, 34:20, are in the ratio of the golden mean, within the limits of accuracy of the measurements. An end-on view of the DNA molecule is shown in a computer simulation in (c). The decagon shape can be visually resolved into two overlapping pentagons.

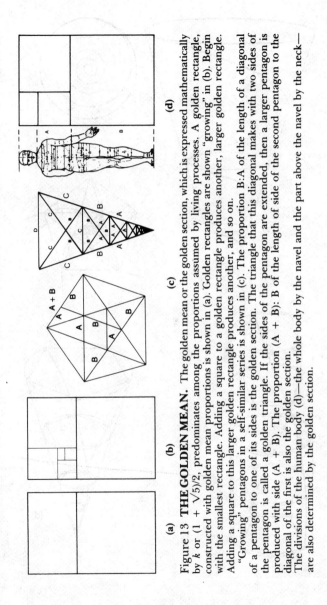

Figure 13 THE GOLDEN MEAN. The golden mean or the golden section, which is expressed mathematically by k or $(1 + \sqrt{5})/2$, predominates among the proportions assumed by living processes. A golden rectangle, constructed with golden mean proportions is shown in (a). Golden rectangles are shown "growing" in (b). Begin with the smallest rectangle. Adding a square to a golden rectangle produces another, larger golden rectangle. Adding a square to this larger golden rectangle produces another, and so on.

"Growing" pentagons in a self-similar series is shown in (c). The proportion B:A of the length of a diagonal of a pentagon to one of its sides is the golden section. The triangle that this diagonal makes with two sides of the pentagon is called a golden triangle. If the sides of the pentagon are extended, then a larger pentagon is produced with side (A + B). The proportion (A + B): B of the length of side of the second pentagon to the diagonal of the first is also the golden section.

The divisions of the human body (d) —the whole body by the navel and the part above the navel by the neck— are also determined by the golden section.

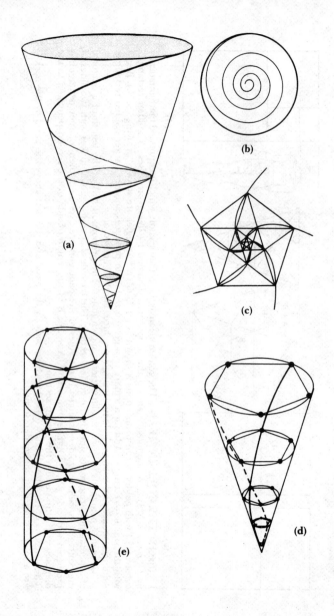

Figure 14 **THE PENTAGONAL GEOMETRY OF DNA.**
A growing negentropic process can be shown by a spiral in a cone (a), which looked at from the top is a simple logarithmic spiral (b). A pentagon whose sides are extended out as shown in (c) will produce five more points that can be joined to create a larger pentagon. Each of the larger pentagons produced this way is rotated one-tenth of a full circle. If this construction were projected and looked at from the top down, it would be a series of pentagons, getting larger, each one rotated one-tenth of a full circle, from the previous one (d). This self-similar series of pentagons in the plane can be viewed as the projection of a conical series with two diagonally placed vertices of the pentagon tracing a double-spiral. This represents the form of the DNA when it is accomplishing negentropic action. Reconverting the conical form into the cylindrical form usually observed (e) gives us a geometrical model of DNA that has the golden section proportions observed.

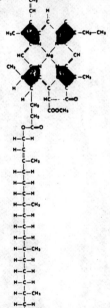

Figure 15 **THE GEOMETRY OF CHLOROPHYLL.** A chlorophyll molecule with four pentagons shaded and pointing toward the magnesium atom at the center, which can be thought of as an "antenna."

Figure 16 **TRANSFORMING THE DESERT.** Man not only has to transform the deserts on Earth, but on Mars. This is a Mariner spacecraft view of the surface of Mars.

LUIS CARRASCO

The Future of Astrophysics

Astrophysics is really so important as a science, because almost any possible physical process that we can imagine, that we could dream of, is likely to take place somewhere in the universe, and is taking place already.

In fact, we should recover our historical background. Physics was started by learning processes that were taking place in the universe; that's what Kepler did. He inferred laws that were controlling the motion of planets around the Sun, and he inferred a process from there; and that is the way science makes progress. And I am putting an emphasis on this point because there is a strong tendency in astrophysics, at least these days, to try to impose on the universe things that we do not really understand. Our rather imperfect knowledge of things, is forced. Somehow we have fallen behind, to a certain extent, and we are making an effort to deny that there are a number of things that are occurring in the universe, which we can observe, but which we don't understand—because we don't understand them, we try to deny that they are taking place! I'll show some examples of astronomical objects, which, as you will see, are very familiar to most of us, and yet which have lots of open problems. Sometimes, we just don't understand some very simple things.

This is a picture of a hydrogen transition on the Sun (Figure 1). The Sun is an astronomical object very familiar to every one of us; we see it every day. On the upper left is a prominence, which is an eruption—an eruptive process on the surface of the Sun. Most of the dynamics of the process of eruption are not well understood. Nevertheless, we try to understand them, and we fail over and over again, as long as we use incompetent theories in terms of trying to understand them. This is obviously a hydromagnetic phenomenon. Associated with that process, there are strong magnetic fields, acceleration of particles to very high energies. In fact, we can detect, with our satellites, the increase in cosmic ray particles that is associated with this type of eruption, x-ray emissions—a whole collection of high-energy phenomena, always associated with this type of process. Well, let me tell you, we don't have a good theory to explain that.

This is the Sun again in Figure 2, and now we see some smaller-scale phenomena. These are always connected with the magnetic activity of the Sun, areas that are preferentially heated, or overheated, in the surface of the Sun. In fact, they by themselves defy what is called the Second Law of Thermodynamics. Thermal theories would predict that the surface of the Sun should be a roughly uniform-temperature thing, except perhaps at the poles, but when we look in close detail at the Sun, we see that it is not uniform, and that it is composed of a lot of vortical structures. These, essentially by having rotational action, combined with electrical charges, then automatically have the appearance of magnetic fields. As mentioned by Jonathan Tennenbaum, these are essentially least-action units, which work for themselves and are long-lived processes. They cannot be treated just as simple perturbations of the process, but in fact they define the process itself.

Here is another example that is also a familiar object to most people (Figure 3). This is the Crab Nebula, which is a star that exploded about 1,000 years ago and was recorded by Chinese astronomers in the year 1054, when it exploded. When we look at the sky in that place now, what we see is a mass of very hot plasma, expanding at very high speed—about 10,000 kilometers per second—and all the structural features that we see, the rings, the filaments, are apparently magnetic, self-confined plasma structures, probably similar to some of the filaments that Jonathan Tennenbaum was showing from the plasma focus fusion device.

Another problem to deal with is the structure and morphology of galaxies. What is a galaxy? A galaxy is a compound of stars (Figure 4). But stars are being formed all the time, and the geometrical site at which new stars are being formed all the time is a spiral structure. As to why this is, well, we have some idea as to what is happening. The gas which is vital—it is the raw material to form new stars—

is piled up, swept, and compressed into a spiral-shaped structure, where magnetic processes are also very efficient in actually providing the proper physical conditions for stars to form. The details of the physics that we can learn by studying this sort of process certainly should enlighten our concepts for any future local application of universal physics.

Here's another galaxy, in comparison (Figure 5), an elliptical galaxy, which is different from a spiral galaxy. What has happened here, essentially, is that the star formation process in the past was so efficient that most of its gas was transformed into stars at a very early stage. We have some evidence, in fact, that this is the basic difference between spiral and elliptical galaxies—the rate at which these different objects consume, or transform, their gas into stars.

A New Physics

What's the future of astrophysics, in terms of space programs, in terms of what should be done in the near future?

First of all, we need space telescopes, for several reasons. As you can easily imagine, Earth's atmosphere—which contains oxygen that is vital for life—is not so good for astronomy and astrophysics. It essentially blurs the image. The information we get from outer space becomes distorted when it gets across the atmosphere, and this distortion blurs the image: We get a foggy picture of what the universe really looks like. Perhaps the most spectacular and important results in the near future will come from apparently very small structures, containing very high-energy-density phenomena that, due to this blurring, we cannot properly study from ground-based observatories. So, one thing that we should have—and we will have very soon—is the Space Telescope.

However, that's only one experiment, one observation instrument, and one is too few.

It figures that if every competent astrophysicist in the world would ask for observing time, each would not have more than a couple of hours within the next five or ten years of operation of the satellite. So, certainly, we need more Space Telescopes. And, of course, the moment we have at least two of these telescopes in orbit, then we can begin to think about what is called aperture synthesis, which is the following. We can place two telescopes in orbit, look at the same object, and by playing certain tricks, in terms of space resolution, in terms of how small an object we can see—we can mimic a telescope of the size of the Earth! This, in fact, would then allow us to see and study many—very many—of the very small-scale, compact phenomena in the universe, including, the nuclear regions of galaxies, where some very impressive phenomena are taking place.

We know, for instance, that a number of galaxies, at very early stages of their evolution, become extraordinarily bright in their nuclear regions. And just to give you an example of our knowledge so far, we understand, for instance, that in the nuclear regions of galaxies, at one point in time, a region comparable to the size of the Earth's orbit around the Sun, is capable of producing, of liberating, energy equivalent to 1,000 times the entire luminosity of the whole galaxy later on.

Whether or not we need entirely new physics to understand these objects, is still an open question. It is my belief that we do need a new physics. The possible explanations are, in a sense, now taken out of science fiction books. We have to look and make better and more detailed observations of these types of objects, and this will be possible only if we put more telescopes in orbit and do aperture synthesis.

Another need, certainly, is the establishment of astronomical observatories on the Moon, because then we will have all the advantages of ground-based astronomy—in the sense of having the stability, the ability of carrying on observations for very long periods of time, of a single object—without the problems of Earth's atmosphere.

It's a pity: When I was a child, or later on, when I began working in astronomy, about 20 to 25 years ago, I always dreamed that one of the first things that was going to happen was the establishment of an observatory on the Moon. And, yet, it hasn't happened! And it just has to happen, because otherwise, the rate at which we are acquiring knowledge is slowed down.

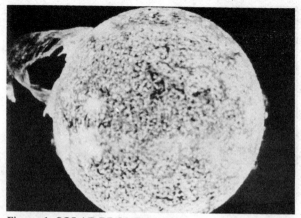

Figure 1 **SOLAR PROMINENCE.** Solar prominences are obviously hydromagnetic phenomena.

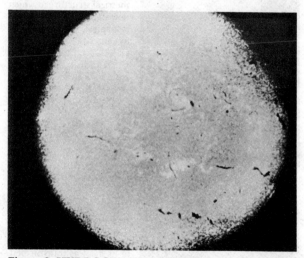

Figure 2 **HYDROGEN TRANSITIONS ON THE SUN.** Smaller-scale phenomena like these are related to the magnetic activity of the Sun and defy the Second Law of Thermodynamics.

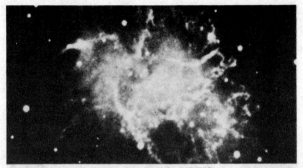

Figure 3 **CRAB NEBULA IN TAURUS.** The filamentary structures in the Crab Nebula in the constellation Taurus are similar to the filaments created in the plasma focus fusion device.

Figure 4 **SPIRAL GALAXY.** The geometrical structure at which new stars are being formed is a spiral. This spiral galaxy in the constellation Canes Venetici is NGC 5194, with its satellite galaxy NGC 5195, known as the Whirlpool Galaxy.

Kitt Peak National Observatory, Cerro Tololo
Inter-American Observatory

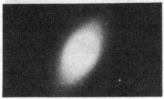

Figure 5 **ELLIPTICAL GALAXY.** Star formation in this elliptical galaxy was so efficient that most of its gas was transformed into stars at a very early stage.

Hale Observatories

DR. FRIEDWARDT WINTERBERG

Changing the Universe According to Our Will

I will speak about the significance of the Strategic Defense Initiative for the future of space flight and fundamental research.

There are two great frontiers of science: first, the quest to reach ever-greater distances in space; and second, the study of ever-smaller dimensions of space and time. Both quests—toward the large and the small (the first done at the moment, still with space rockets, and the second done with high-energy particle accelerators)—require ever-larger energies.

Krafft Ehricke put forward the thesis that man is going to transform his universe—that means our universe—according to his intelligent will. Thus, man will make, among other things, uninhabitable planets habitable, and, in the distant future, eventually he will even mine suns. Krafft Ehricke wanted to change the universe, according to our will—not just to disturb the universe, as it was suggested by Freeman Dyson. However, whether Krafft Ehricke's dream comes true or not, depends on whether we are beset by a new dark age of superstition, astrology, or whatever, which we already can see, unfortunately, these days in the pseudosciences.

Unless human failures take place—which, of course, I cannot predict—I think Krafft Ehricke's dream depends upon another very important conjecture. (A conjecture, of course, is a theorem that we do not know is really a theorem. If we have a conjecture, it *might* be a theorem.) This conjecture, I would like to call "the physical technical principle." It says that the laws of nature have a structure that permits a technology that makes all these laws explorable. I should emphasize that it is by no means obvious that such a theorem exists; it is a conjecture. However, this conjecture, until now, has been proven to be true.

In mathematics, we have conjectures like that famous conjecture of Bernhard Riemann about the theorem of the zeta function, the prime number theorem. Of course, there too, we only know that until now it has been shown to be true; we have not proven that it is true, which means it must be proven absolutely true for all circumstances. However, my hunch is that the Riemann conjecture is probably true. Even though I am not a mathematician, I think that if something has been shown to be true in that many cases, it is hard to believe that mathematics, which is in some way also part of nature, should suddenly abandon us.

Likewise, in our subject here of physics, it is hard to believe that suddenly the experience we have had so far would abandon us, and we would suddenly end up with certain regions that become unexplorable. So I would like to say that everything that can be explored, not only *will* be explored, but everything that is explorable *can* be explored, which is something much different. We can, of course, say, that everything that *can* be explored will be explored, but it doesn't mean that everything that is explorable, can be explored. There may be things where our technology simply doesn't permit it to be explored. And the "physical-technical principle," which is of course a conjecture, says that everything that is explorable can be explored. Therefore, we require a certain technology that makes such exploration possible.

Colonizing the Galaxy

Let's reflect backwards. First of all, that the exploration of the solar system, the exploration of the surfaces of the planets, is possible, is, I think, established beyond reasonable doubt since the successful completion of the Apollo program. There can be no doubt that eventually we will have the technology to explore all the surfaces of all the planets of our solar system. Furthermore, there cannot be the slightest doubt that if we master the technology of controlled fusion, we must only miniaturize it. In itself, I think there is no doubt that we will master it, because actually we have already begun to master it. In the hydrogen bomb, we have mastered the technology of fusion; we must miniaturize it. So there will be no doubt that we will eventually succeed in miniaturizing it, and that we will be able to reach nearby solar systems with fusion propulsion.

Then, if we go, let us say, to a nearby solar system maybe 10 light years away or maybe 100 light years away, which is habitable, and if we then stay in that new solar system to colonize it for a thousand years, and then keep going on, hopping from one solar system to the other—a simple calculation shows that the entire galaxy in this way could be colonized in about 10 million years.

However, I will not go into this because it would be the subject for a separate lecture. Let me just say that the exploration of our solar system, of nearby solar systems, and finally of all solar systems in our galaxy, is clearly feasible with fusion propulsion.

I would like to raise arguments that support the physical-technological conjecture I had, because only if this conjecture is true, will we really become masters of our universe. We are not just interested in exploring surfaces and planets and colonizing them; we would like to know—using a phrase by Goethe— the inner forces of nature that are holding together the inner parts of nature. That means, atomic and subnuclear atomic physics, but also biology. And here is now something very important, where the SDI, the Strategic Defense Initiative, enters. Because the implementation of the SDI will give us new tools, which represent what I may call a kind of super-technology, some kind of a super-technology, as yet *unheard of*. And it will therefore very well give us the tool, eventually, to explore all laws of nature. And that's what I would like to talk about here in my lecture.

Therefore, I would like to review some of the ideas of the SDI; and I would like to give a short status report on the SDI—at least on those concepts of the SDI that right now indicate how such a defense system actually could be built in a relatively short period of time. However, before I give you such an outline, I would like to reflect back to some ideas Hermann Oberth had in 1924, in a letter he sent to a publisher for a planned book, in which he outlines some of the ideas that, actually, now are part of the SDI.

I will translate Oberth's letter from German: "Here is another problem: is it perhaps possible," Oberth asks, "that one can propagate, in space, beams [he thought about cathode rays or ion beams] which remain parallel over arbitrary distances?" The laser was not invented at that time, so Oberth thought about particle beams. He continues: "Such beams, then, could serve, in a certain sense, as tracks in space, which could be used by spaceships to receive energy, and also could be used to give the spaceships some kind of a material hold."

Oberth also notes, "It is quite conceivable, in fact, that one could use such beams as tracks—for example, between two parallel beams, to make the analogy complete, of railroad tracks." Oberth thought in terms of positive and negative charged particles, which of course would be necessary if the particles are charged. In the case of an uncharged laser beam, we would only need one beam, one track; one is a monorail, and the other one is not the monorail!

Then Oberth raises the question, if a spaceship were helped by such beams, could it attain the velocity of light, or come close to the velocity of light? Thus he raised questions which have attracted, decades later, science fiction writers. If then Einstein's special theory of relativity would in fact retard, as seen from the earth, the processes in the spaceships, then everything would go slower. Oberth expresses the view that there cannot be the slightest doubt, in spite of many other views, that Einstein was absolutely correct. Before the time of Einstein, dilation was actually observed. As we know today, the time dilation effect is true.

Now let me review some ideas on the status of the SDI.

One very important breakthrough in the SDI was made a few years ago—it is what we call the x-ray laser. In a laser, you need some kind of energy source to pump some medium. The medium consists of atoms, which all oscillate in a coherent way and emit a beam in one preferred direction only. I have not the slightest doubt that the x-ray laser concept developed by Lawrence Livermore is a direct thought-child of the famous Teller-Ulam configuration, or I should more accurately say, the Teller-Ulam-Guderley configuration. In the Teller-Ulam configuration, Teller combined ideas, unrelated ideas by Ulam and Guderley, into the famous configuration. Teller invented the configuration.

The Teller-Ulam configuration appears in Figure 1. You have an exploding atomic bomb, even though it's an ellipsoid. It can be an egg-shaped configuration. Now we have a U-235 fission explosion; actually, the diameter of that deuterium ball at the right is about 1 meter, and the other thing is very much smaller. At the moment when the U-235 chain reaction is going up, an enormous amount of soft x-rays is emitted from that explosion.

Now here is Ulam's idea. These x-rays can be confined by solid walls so that the x-rays flow in this cavity, and then hit the deuterium ball on the right from all sides. What you see as GSW is a Guderley shock wave. This is a famous solution by the aerodynamicist Kurt Guderley, who by the way, is one of the Operation Paperclip people not yet asked to return to Germany like Dr. Rudolph was. Guderley also got the highest U.S. award, the Distinguished Service Medal. There's a very famous theory by Guderley: If you have a convergent shock wave, you get about a 30-fold compression in the center, and the temperature rises from 10 million degrees to 1 billion degrees, in this case of a meter-sized sphere. Then you get ignition in the center, and a thermonuclear propagation wave moves outward.

The hydrogen bomb, which was exploded in the Bikini Atoll, called the "Mike" Test, must have looked like that.

The x-ray laser is shown in Figure 2. We deform Figure 1 in such a way that it becomes a long pipe. Inside the pipe is a wire. That wire is the laser rod; it's bombarded with the x-rays, and it becomes, incidentally, what laser physicists call a continuous wave laser or CW. The laser transitions in the x-ray domain are very, very short-lived, but the x-ray bombardment before the whole thing disintegrates with an explosion wave, lasts long enough to pump that many, many times up and down, so that we get a beam of x-rays.

More recently—we do not know, but we read some reports in the *New York Times*—a breakthrough has apparently been made by Lawrence Livermore National Laboratory which, if it's true, amounts to the fact that a certain sort of an optical cavity has been made with such x-ray lasers. If that is true, then the beam reaches such coherence that it doesn't spread out anymore. In this type of configuration, the beam would still spread out, because the pipe, or the wire, is not infinitely thin. It has a finite thickness, and, of course, if I were to make it too thin, then no radiation would come out. So the spreading of the beam depends on the ratio of the diameter of the wire, divided by the wavelength. So the beam actually spreads out slightly; over several thousand kilometers, the spreading-out can be quite significant.

If the breakthrough reported in the press were true, it could very well be that they succeeded in making an optical cavity. If that's the case, then we not only can make the x-ray laser beam much, much stronger, but we can also keep it together, and that of course will have tremendous implications for Oberth's idea. If we can really make such an x-ray beam and keep it together, then we can propel a spacecraft over very large distances, even making interstellar flight feasible.

I would like to point out, as an aside, one other very important application of the x-ray laser. As you know, there are tremendous efforts under way to miniaturize hydrogen bombs to such an extent that the explosion can be controlled in a chamber, thereby, replacing the fission trigger by laser or particle beams. In other words, we would have some kind of an explosion motor, where miniaturized explosions would take place (like the explosions in an automobile engine, which do not destroy the automobile engine; they are controlled). We need, of course, a different kind of trigger. If we have that, we can have what you call thermonuclear microexplosions.

Sandia National Laboratories has made a tremendous breakthrough just recently with these ion beams and a magnetically insulated diode. They can now focus these ion beams to a spot less than a millimeter. It is almost like an optical lens. It was totally unpredicted, and it is theoretically not yet understood, because the thing is highly nonlinear, but by trial and error, they found that out. Of course, that is a very big challenge to Livermore, because Livermore only has a laser, and now the Sandia people have a megajoule ion beam machine—so they are by overall effect 10 to 100 above Livermore at least. This ion beam machine gives them much more energy.

Spinoff Applications

Well, if we can do that, then we can also make an x-ray laser for the laboratory. And then, by a certain technique called holography, we could magnify living tissue, living cancer cells, a million times, and the cancer cells would not be killed. In an electron microscope, in contrast, the cancer cells being looked at are killed. So we could make enormous magnifications, because x-rays are much less destructive than the electron beam rays in an electron microscope. If we have magnified a cancer cell as big as this room, then we can see exactly which molecule is moving from here to there, and we could have a very clear idea of the cause of the abnormality in cancer. We could explore the cause.

Of course, cancer would not be the only thing; the x-ray laser would also apply as well to other kinds of diseases. And in genetic engineering, we could really see that we must move "this piece" from over here to over there. In fact, we will then eventually be able, I should point out, as a side-project of the SDI, to read the genetic code like reading a book. And then, of course, since our life lasts— what does the Bible say?—70 years, and now it's reaching the 80s. (But, of course, I know people like Oberth who is 91.) So, we need about 30 years to get educated, and then another 30 years we work, and then we are thrown away.

How long does it take today to produce a car? A day? (Well, we don't produce anymore, but maybe the Japanese do.) What if the car can then move around for one day, and then it goes to the junk dealer? This is obviously very inefficient. Since our life is very short, for evolutionary reasons, I would like, with genetic engineering, to change our lifespan so that we get maybe a thousand years. I would like personally, if I could, to live forever! But at least, if we could, with genetic engineering, increase the lifespan—of course that means to retard the aging process—by a factor of 10, that would be a breakthrough in everything. Our education, which lasts 30 years, could be much more efficiently utilized.

So you can see what kind of side effects the SDI may have, side effects that are very difficult even to predict at this time.

Interstellar Travel

There is another very important milestone in the SDI—a magnetic accelerator (Figure 3). The magnetic accelerator could be driven by thermonuclear microexplosions. There are several chambers where small nuclear explosions take place; each chamber produces a fireball that presses against the magnetic field, and an enormous electromagnetic pulse is produced.

We can launch very large payloads from the surface of the Earth into Earth orbit. Furthermore, since we have to go through the atmosphere, we could use particle beams and laser beams, first to make a channel through the air, heating it up to several thousand degrees; then the air will become much thinner. And maybe we can go up to 10,000 degrees, even 100,000 degrees. Of course, the channel would only have to last for a very short moment, then the projectile would go through that channel, and would have practically no air resistance.

In Figure 4, we see how that works in the framework of the SDI. We would have huge electromagnetic guns, let me say, in northern Canada, if the Canadians would like to participate with us, or Alaska, or some of these Arctic islands. The problem is that if there were a launch of boosters, of rockets, from the Soviet Union, in that case the x-ray laser must be lifted very rapidly above the horizon. Ordinary rockets may not be fast enough; but if we can make an electromagnetic gun where the projectile moves at, let us say, 30 kilometers per second—which I hope can be attainable—then, of course, in less than a minute, or maybe a hundred seconds, the x-ray laser would be very high up. Then it can be fired, and can hit the boosters.

Now, there has been a lot of talk that so-called fast-burn boosters could escape the x-ray laser, because fast-burn boosters are rocket boosters that reach a very high velocity very fast. The MX is already a sort of a fast-burn booster. The Russians do not have a fast-burn booster; the moment when the whole payload of warheads and decoys is released is relatively high—I think roughly 200 kilometers above the surface, if I am not mistaken.

But a fast-burn booster can already reach that velocity at 80 kilometers altitude; and then, the x-ray laser would be absorbed. The x-ray laser beam has some spreading out, but now Livermore apparently has succeeded in making a sort of an optical cavity with the x-ray lasers. Then the laser would simply burn a hole through the atmosphere and could destroy the fast-burn booster. So I think this idea with the x-ray laser, in combination with an electromagnetic gun, has to be taken very seriously. We can also use an electromagnetic gun as a zero-stage for a rocket, and there may be some combinations that will turn out to be optimal.

In Figure 5 I demonstrate some idea of how we can use the electromagnetic gun technology to reach interstellar distances, maybe a trip to a nearby solar system. What is suggested here, is the following: We have an electromagnetic gun here again—an electromagnetic gun with many capacitor coils. Of course, we could also use thermonuclear microexplosions instead of these capacitors. Suppose we shoot an electromagnetic gun before a spaceship starts. We would project a whole chain of nuclear explosions. This chain is now sort of a road, a beam, but it's moving ahead of the rocket. Then the rocket comes like a ramjet, and whenever there is a nuclear explosive, it's ignited with a shaped charge, and it gives the space ship a push.

So if we make a very, very long road of such nuclear explosives, and the spaceship comes and sets off the nuclear explosive cells, the nuclear explosive will be detonated behind the spaceship. To attain velocities of 3,000 kilometers per second, we can make a shaped charge, a thermonuclear shaped charge, as shown in the figure. For very large velocities, we make a different kind of explosion, which would produce velocities close to the velocity of light.

Unfortunately, I cannot discuss how these nuclear shaped charges work and why one shaped charge would produce less and the other more; with one you get maybe a three-times-higher velocity than the nuclear explosive products move. In a thermonuclear explosion you have about 10,000 kilometers per second versus 30,000 kilometers per second; and in the other we can come close to the velocity of light, especially if we have a very narrow angle. We would like to get to velocities maybe 90 percent the velocity of light. We would have to shoot many, many such atomic bombs, or rather hydrogen bombs, ahead of that spaceship, and that spaceship, would use all of the explosives already placed there, very nicely, as a sort of a rail. So the energy, and the recoil mass are already put ahead of the spaceship.

A Gamma Ray Laser Beam
to the Stars

Now, there was a discussion about a nuclear ramjet using interstellar gas. That is not going to work, because the density of interstellar gas is much too thin, and there is also not enough thermonuclear fuel. But of course, here, there is one possibility for making such a road; I call it a nuclear road to the stars.

In Figure 6 I come to something extremely exciting—a new concept that I hope very much will be a new kind of area of research. I should point out, that there has been lately, a renewed interest in an old idea that had been first described by a physicist who is now quite old, Willard Bennett. Bennett invented the pinch effect in 1930 or so. The idea of using the pinch was popular in the early days of the fusion program. One pioneer on the pinch research, Dr. Winston Bostick, is sitting here, and he still believes in the pinch, and I think a lot of people are starting to believe again in the pinch, in an entirely new light.

I would like to describe to you in a moment, a very, very unusual kind of pinch, for which we have to, however, first produce a very large number of positrons. One researcher, a Russian researcher with the name Meierovich from the Soviet Academy, says in an article he published not too long ago, that he believes the entire future of physics will depend on this sort of pinch. I will briefly, describe what the pinch idea is. First of all, we need a very large number of positrons. We will need a plasma that does not consist of electrons and protons, for example, but of matter and antimatter. Matter and antimatter, in this case, an electron-positron plasma. Now, in order first to make this plasma, we need to produce a very large number of positrons. Positrons are electrons except that they have an opposite charge. And if one meets an electron, they annihilate each other into gamma radiation.

In the first step, we can produce a very large number of positrons with an ultraviolet laser. In the second step, we would then accelerate the positrons to relativistic energies, and make them form an intense relativistic positron beam; then we would do the same thing with electrons. Of course, we would not have to produce the electrons, because they are available. In the third step, we would make the electron and positron beams coalesce into a relativistic electron-positron plasma of counterstreaming beams.

Suppose beams would have to be accelerated into opposite directions, for example, into large rings or racetracks. They can be many, many kilometers long. One beam goes in one direction and the other goes in the opposite direction. And then both beams are oppositely charged, so they attract each other electrostatically, but also magnetically; they would coalesce in one electron-positron plasma. The electron-positron plasma would then first be neutral, because the positive charges of the electrons would be compensated by the negative charges of the positrons-electrons. Opposite charges compensate each other; and we get such an electron-positron plasma as shown in Figure 6(c).

A positron beam is shown in (a); (b) would be an electron beam; they are coalescing in (c). And now something extremely interesting happens. In a paper I published in 1979 in the *Physical Review*, I showed that such an electron-positron plasma would have two effects. First, if the particles move toward each other, they collide. This collision of particles will result in heating—so the beam gets hotter, and the beam would blow up. But there is another effect because the beams are relativistic. These collisions produce oscillations perpendicular to them, and that plasma, the electron-positron plasma, loses radiation. There are some situations where it can lose more heat and lose it faster than it is produced. Then, of course, this electron-positron plasma will collapse. The question is, to what diameter will it collapse?

The answer is, it can collapse to a diameter of 10^{-13} centimeters—to nuclear dimensions—or, under some circumstances, to even a smaller diameter. So we would have here an entirely new state of matter, which in the universe exists only in neutron stars, or in stars that are closely related to black holes. In neutron stars—we have densities as in the nucleus. We have a radiative collapse of a relativistic electron-positron plasma. First of all, just before the collapse, this filament, which is fantastically narrow, would send out an enormous burst—not of soft x-rays but of gamma-rays—that means very hard gamma rays.

Figure 7 shows—from my *Physical Review* paper—how we could produce such electrons and positrons in some sort of a modified betatron, since two beams would be produced and then be pushed together, forming the plasma. The positron beam and the electron beam would be pushed together, forming the plasma. Of course, instead of making this plasma, which may have here a dimension of meters, I could make it in a large accelerator like in a racetrack, many, many miles long, to get much, much more energy. If I can produce enough positrons, there is no doubt we can do that with lasers.

Now, what we can do? We have a many-mile-long racetrack-like relativistic electron-positron plasma, which would produce a gamma-ray beam of fantastically high coherence, because it goes down to 10^{-13} centimeters—fantastic coherence (Figure 8). But it would have an enormous, very large, energy release, because it's a pinch of nuclear density. So we could use this gamma-ray beam to give energy to an interstellar spaceship, as Professor Oberth had envisioned it. We could now project such gamma ray beams produced in this way over very, very large distances, because the beam is so highly parallel and highly coherent, that it would not even spread out significantly over interstellar distances. So, besides the nuclear road to the stars, we have a gamma-ray road to the stars, a gamma-ray laser beam to the stars!

Now let me return to my thesis, or conjecture, that the laws of nature are made and have such structures that we can explore, that we can make a technology that permits us to explore the laws of nature. We have seen in chemistry what we can make with the relatively small energy of electron volts. Chemistry was definitely a great advancement. In order to grow to megavolts, we needed special accelerators, and megavolts opened us to nuclear reactions.

With nuclear reactions—I must come back to

Einstein's equation, $E = mc^2$—nuclear reactions release only 1 percent of the energy in matter. Fusion releases somewhat more than fission, but we do not get out 100 percent of the energy. Now, if we go to much higher particle energies, if we have accelerators producing much higher particle energies, we may get into a situation where the decay of protons can be accelerated. People are presently wondering if protons decay. So far the evidence is not conclusive, or even negative, but, of course, that doesn't mean they cannot decay. Radium decay was not easy to detect, and required certain very delicate measuring equipment. It is quite conceivable that protons also eventually decay through an interaction that only becomes significant if you have reached extremely high energies, several MeVs.

For this, we need gigavolt energies, not just the megavolt energies typical of nuclear reactions. There is some speculation that the energy of strong/weak electromagnetic interactions becomes one such interaction. And there is another energy which is even 100,000 times higher. With such electron/positron plasma nuclear states, we can think of exciting relativistic electron/positron plasma nuclear density waves; and it is quite conceivable that with the help of such waves, we can make some sort of laser accelerator. Now, however, the acceleration takes place with gamma rays, in that very highly dense matter. Enormous particle energies could be reached. In fact, we can reach particle energies that otherwise, with present-day accelerator technology, could be accomplished only with an accelerator many light years long, which definitely would not be technically feasible.

With this highly condensed state of matter, we could conceivably reach particle energies. So what Meierovich said, that the entire future of physics may depend on that, is quite a well-justified statement. We get nuclear densities in the laboratory, and there are some interesting applications for short-range use. That is what the Europeans are particularly interested in. If you could make such devices in Europe with the velocity of light, these intense x-ray beams, gamma-ray beams, could puncture through the atmosphere exactly as a bullet does.

Jonathan Tennenbaum, Chairman

I would like to add something I omitted from the introduction. Professor Winterberg studied with the great Werner Heisenberg, the German nuclear physicist who was one of the fathers of nuclear energy. It is interesting to note that Krafft Ehricke, during the war, actually spoke to Heisenberg, who was developing a nuclear reactor. He got the specifications and began to design in Peenemünde, as part of the group there that was working on advanced space projects, nuclear-powered rockets. In fact, if you look at documents of plans, detailed plans that the Peenemünde group turned out during the war, there is very little new that we've done in the NASA program—including the NERVA program, which was unfortunately abandoned, of developing a nuclear-reactor rocket engine. There is very little that wasn't already at least in the first approximation design in the Peenemünde papers.

I would also like to point out that if you look at the science we've been discussing this morning, it should be very evident that this science is international science, that many of the crucial contributions to the breakthroughs we are talking about, have occurred just as much in Western Europe as in the United States. That's one very crucial aspect of what we're going to be talking about this afternoon; namely, the cooperation between the United States and Western Europe on the SDI. Professor Winterberg has already mentioned it, in terms of the practical defense of Western Europe, but it must be emphasized again and again that equal-partnership cooperation between the United States and our Western European and Japanese allies will be absolutely decisive, and the brain power, as indicated by their scientific breakthrough, will be the driving force.

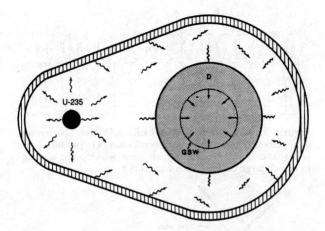

Figure 1 **TELLER-ULAM CONFIGURATION FOR AN ATOMIC BOMB.** An enormous amount of soft X-rays are emitted in the U-235 chain reaction.

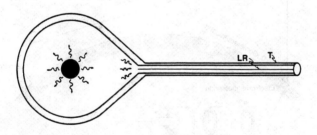

Figure 2 **X-RAY LASER CONFIGURATION.** Deforming the Teller-Ulam configuration into a long pipe with a wire inside, produces a continuous wave X-ray laser.

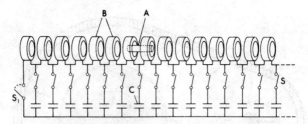

Figure 3 **MAGNETIC ACCELERATOR.** Thermonuclear microexplosions in several small chambers produce fireballs that create an electromagnetic pulse, which could be used to launch large payloads into Earth-orbit.

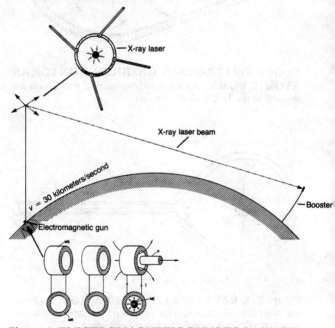

Figure 4 **ELECTROMAGNETIC GUNS TO LAUNCH X-RAY LASER SYSTEMS.** Huge electromagnetic guns in northern Canada could pop-up X-ray laser systems at 30 kilometers per second to kill Soviet rockets in the boost stage.

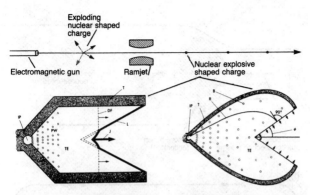

Figure 5 **A NUCLEAR ROAD TO THE STARS.** In this configuration an electromagnetic gun projects a "road" of nuclear explosions into space. The spaceship detonates the explosions, which then propel the vehicle forward.

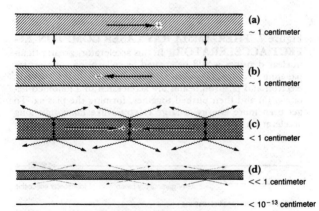

Figure 6 **SCHEMATIC FOR RADIATIVE COLLAPSE OF AN ELECTRON/POSITRON PLASMA TO NUCLEAR DIMENSIONS.** In this concept, a large number of positrons are created, accelerated to relativistic energies, and coalesced with a similar beam of electrons into a relativistic electron-positron plasma of counterstreaming beams. This beam will collapse to a diameter of nuclear dimensions, or perhaps smaller, creating an entirely new state of matter.

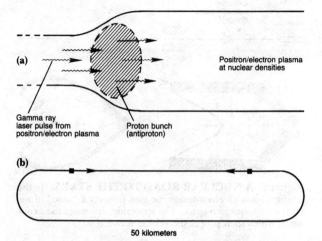

(a)

Positron/electron plasma
at nuclear densities

Gamma ray
laser pulse from
positron/electron plasma

Proton bunch
(antiproton)

(b)

50 kilometers

Figure 7 **THE GAMMA RAY LASER COMPTON EF-FECT ACCELERATOR.** In this accelerator, greater-than-nuclear densities would be created by collaspsing counterpropagating electron and positron beams together. In this concept, electron and positron beams are accelerated in a modified betatron (a) and then pushed together, forming the plasma. To get even more energy, a large racetrack-like many-mile-long accelerator could be used, such as in (b).

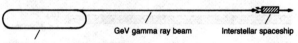

GeV gamma ray beam   Interstellar spaceship

Relativistic electron/positron plasma

Figure 8 **GAMMA RAY LASER ROAD TO THE STARS.** A many-mile-long racetrack-like relativistic electron-positron plasma would produce a highly coherent gamma-ray beam of enormous energy. This beam could then be used to give energy to a interstellar spaceship, as professor Oberth had envisioned it—a gamma ray road to the stars.

The Challenge of Creating Operational Beam Defense Systems Within Three Years: The Role of Cooperation Between the U.S. and Its Allies

Afternoon Panel
June 16, 1985

HELGA ZEPP-LAROUCHE

What Is Wrong with the Western Alliance?

What kind of shape will the alliance take with the new doctrine of Strategic Defense Initiative (SDI) and Mutually Assured Survival? The reason why the alliance and Western civilization, and therefore the freedom of the people of the Free World, are endangered in the first place, is that the doctrine on which NATO has been based for the last 20 years, the doctrine of MAD, Flexible Response, deterrence, has a flaw in it from the beginning. This is the idea that war would mean the total destruction of a potential aggressor. The idea that mankind can only survive and that peace can only be maintained because missiles are directed against each other's nations, cannot be what the meaning of mankind and the ideal of man has to be.

And therefore, when President Reagan promised the change from MAD to Mutually Assured Survival, to develop through the new defensive beam technology the means to destroy nuclear weapons and make them obsolete forever, this laid the groundwork for reaching a more reasonable condition of man, to create the kind of condition where the Age of Reason can be possible.

The only problem is, we are in a prewar situation. Look at the immediate crisis that the U.S. government finds itself in today, with the TWA plane hijacked, where now in Beirut more terrorists have gone aboard with grenades and other weapons, threatening the lives of the passengers. The treasonous press has leaked out the fact that the special counterterror "Delta Force" of 50 people, has been deployed to do something about this—virtually guaranteeing that such an operation would be doomed to fail.

This is one of the many incidents which mean that the United States must answer with absolutely complete certainty, that it is not going to take this. Washington must deliver a clear signal to the Soviets, who, in the final analysis, are behind this, through Syrian intelligence and through Iranian terrorists. This is the kind of prewar Sarajevo situation in which we find ourselves.

But despite this, and despite the understandable fact that tensions are high, we still have to formulate a policy which permits a change, a fundamental change in the alliance. The problem was that the founding principles of NATO which have operated in the entire postwar period, were based, not really on the principles for which the American Revolution was fought—for example, the principle of national sovereignty. The key problem of the United States and its allies has been that there was this superpower game, in which each nation in the world was drawn into the sphere of influence of either the one or the other power, and while this may have been halfway possible in the periods of relative prosperity in the '50s and '60s, right now we are in a situation where, because of the world economic crisis, the United States is about to lose all its allies, because it is governed by the wrong economic policy.

One of the things which has become apparent in the many discussions we have had since this Krafft Ehricke Conference started, is that especially the representatives from the foreign countries, from Europe, Asia, Latin America, and so forth, basically want a change in U.S. foreign policy toward them. They want to be treated as equal partners; they want their national sovereignty to be respected, and for the United States to change its policy so that the best interest of each of them is respected. And can you blame them? Because the nation-state is the most important fundament for the values of what we call the free world. How can you be a patriot if you cannot be part of your nation, how can you be expected to defend your nation if you are not permitted to be proud?

The problem has been that, because of the Eastern Establishment, and because of traitors in various administrations, including this one, there has been an attitude of treating the friends and allies of the United States as *assets,* as people you can go to and dictate terms and say: Okay, you're our asset, we tell you to do this, we tell you to do that. And you know, because of the power combination and because of the historical realities—including black-

mail dossiers and this more ugly side of politics—
that it has functioned.

But right now, the Soviets are trying to decouple
Western Europe from the United States, and I
hardly need to repeat: Look at Greece, Spain, Por-
tugal; look at France; look at Germany; look at the
Scandanavian flank, and you see it's hanging by a
thread. We must change the alliance and base it on
those principles that were the reason why the re-
publicans of Europe helped the American Founding Fathers to install on these shores a new nation,
which for the first time in history was a true republic, in which each citizen was as God has created
him—namely, equal—and there were no oligarchs,
no upper elites, no monarchies, no princesses, but
each citizen could be an equal partner to the next,
and a republican.

This principle, if it is good for the United States,
can only be good for the rest of the world too.
Because it lifts our eyes to the real task of mankind,
into the exploration of space and such tasks of the
future, the SDI—or more precisely, the Mutual
Assured Survival doctrine—has to mean also a
change in the relationships among countries. The
Europeans are demanding: "We do not only want
to produce isolated components for the SDI, we do
not only want to be used as assets, but we want to
be equal partners of the United States." And if
Japan demands the same thing, and if developing
countries come—as they have come to this conference—and say, "We may not have the full range
of all the technologies, but we can contribute in this
particular area and we want to do so," then I think
is is very important that this conference send a very

strong message to the patriots of the United States who are right now rallied around President Reagan and Defense Secretary Weinberger, to strengthen them against those people who are in cahoots with the Soviet Union in trying to pull the economic rug out from under the SDI.

So what we are really talking about this afternoon is, what kind of changes have to occur, what kind of philosophical changes have to occur in the relationship among nations, in foreign policy attitude.

For a Second American Revolution

We are right now in a crisis, culturally, strategically, economically, probably worse than at any point in history, probably worse than the Middle Ages, probably worse than the fourteenth century, when plagues, Bubonic plague, and other plagues killed two-thirds of the population in the area from India to Ireland. This is happening in the developing countries today. And the moral crisis: If you look at the punks and the drug addicts and other degenerates, why then the witches of these periods were relatively sane people, compared to what is "normal" social life in some of our countries today. Therefore, we have to have a new just world order in which the national sovereignty of each country is the basis for a community of principle among equal partners and equal nations. But this can only happen, if at the same time a true second American Revolution is carried out, as President Reagan was promising, a real renaissance, like the Italian Renaissance, which in the fifteenth century was laying the foundations of cultural developments for 500 years to come, or the like the German classical period. We need a transformation of people as well, because if the people remain as they are, if they are not willing to change, I'm afraid we will not survive as Western civilization, and apart from the practical, technological, and foreign policy questions, I think the moral challenge to all of us remains actually the biggest one.

So with this, I want to open this afternoon panel. I want to first invite a statement by Gen. J. Bruce Medaris to be read, or rather, it is on a tape which he sent to us, since he was not personally able to come to this conference. General Medaris, as you know, was the fulcrum of the Huntsville Redstone Arsenal for many years. He was a person who did a lot for scientists like Mr. Dannenberg, who spoke yesterday, and I want to invite Jeff Steinberg [counterintelligence editor, *Executive Intelligence Review*] to give an introduction to the tape.

Jeffrey Steinberg

Thank you, Helga. General Medaris asked me to extend his personal greetings to the conference and to communicate his disappointment that he was unable to be here today. When I described to him yesterday the diversity of attendance and the wide range of themes that have been discussed here, he was tremendously excited, because those are precisely the issues which have been the burning passion of his entire adult life. From the period of 1956 through his retirement from the U.S. Army in 1960, he was the commander in chief, first of the Army Ballistic Agency at the Redstone Arsenal and, beginning in 1958, he was the director of a broader program, the Army Ordinance Command, which placed him in charge of the national infrastructure of U.S. Army missile projects, including the Jet Propulsion Laboratory at Cal Tech University.

In this capacity, it was under his direction and through the efforts of the von Braun team that the United States launched the first Earth orbital satellite and some of the very successful early missile programs, and through which the United States developed the first antiballistic-missile defense system, the Nike-Zeus. On his retirement in 1960, the general wrote a book called *Countdown for Decision,* in which he warned, even before the era in which the McNamara whiz kids came in and destroyed U.S. military capabilities through the disease of systems analysis, that he already saw the signs of an impending disaster. He called at that time for a mobilization very much along the lines of that we have effected here.

Yesterday, in the course of our discussion and taping of this statement, he recounted to me one of the most telling experiences he had during his entire period at the Redstone Arsenal. As the director of the Army Ordinance Command, he insisted on and received the authority to maintain a totally independent command over the deployment of all the resources at his disposal, and also the authority to pluck out particularly talented individuals from any other sections within the U.S. Army.

This was the kind of priority that this program was given at the time. In 1958, he took the opportunity to put together a special team of people, centered around the group of German-American rocket scientists who were already there at Huntsville, and closeted them away for two weeks, as the result of which they produced a six-volume proposal which detailed the means to achieve the creation, construction, and manning of a scientific colony on the Moon. He proposed to begin in 1958, and had a time deadline which he was confident could have been met, that would have accomplished that project and set the first scientific team into operation on the Moon eight years later. General Medaris indicated that, given the capabilities of the arsenal system as it existed in the Army at the time, this would have been accomplished on the basis of alternative uses of technologies that had already been developed, and it would have been accomplished on a budget significantly under that which was later allocated for the Apollo Program that accomplished the first Moon shot.

The document was presented to the Secretary of the Army, who immediately classified the study and explained to General Medaris that he did not dare pass this on to the Secretary of Defense, because at that moment the Army was in an intensive battle over who would get the contract to develop a particular missile system, and that if he had dared to submit this as a proposal in addition to the missile program, they would be in danger of losing everything. So I think this was a sort of prophetic case study, from a period in which things had really not gotten as bad as they are now.

Without any further introduction, we should let General Medaris communicate his own thoughts to the conference.

**GENERAL
J. BRUCE MEDARIS**

Stop the Assault Against German-American Scientists!

I cannot believe that the law under which the Office of Special Investigations (OSI) of the Department of Justice is operating, to harass a particular group of naturalized citizens of this country, is constitutional.

Unfortunately, those being harassed[1] do not

---

1. As a result of a witchhunt against German-American rocket scientist Dr. Arthur Rudolph by the OSI, he was blackmailed into renouncing his U.S. citizenship and returning to the Federal Republic of Germany. Dr. Rudolph was the inventor of the Pershing missile and had received three government awards for his contributions to U.S. military and scientific advances. Dr. Rudolph was targeted by forged Soviet documents and perjured witnesses. Maj.-Gen. Medaris sent a letter to President Reagan on May 24, 1985, co-signed by 108 members and former members of the U.S. rocket and space program, requesting White House action to restore Dr. Rudolph's citizenship.

command the resources to challenge these procedures in court, let alone to carry the case to the Supreme Court, which would probably follow. To the best of my knowledge, no charges have been stated in specific terms and supported by proper, direct testimony from identifiable witnesses, and the ordinary rules governing the deportation of aliens have been completely ignored. The individuals targeted for harassment as war criminals are called *aliens,* when, in fact, every condition governing immigration and naturalization was complied with long ago.

Even the public statements of the OSI and other protagonists of these unjust actions have alleged guilt by association. Direct evidence supporting the innocence of Dr. Rudolf and others now under attack are totally ignored. To allow continuation of this kind of treatment, smacking strongly of blackmail and the lawless methods of underworld enforcers, is to ignore such a travesty of justice, that makes a mockery of the Constitution and of every so-called civil-rights law aimed at equal standing before the law for all the citizens of this nation.

While the particular assault on these basic values, to the defense of which I have devoted most of my life, is most urgently before us at this time, there are many other challenges that demand the attention of all intelligent patriots. I invite you to consider just a few:

First of all, the obsession with Mutually Assured Destruction has so controlled our use of resources as to bring about some dangerous and disreputable conditions in our presently available forces usable in conventional warfare. This, in the face of the fact that non-nuclear warfare has repeatedly challenged our national interests.

Second, success in bringing about an effective strategic defense against a possible nuclear strike will bring into sharp focus the threat of conventional warfare, and I feel confident that at the present time the forces opposed to our way of life, total much more than our own and those of our potential allies.

Third, it is worth considering that, since the time when in Korea the advice of the military commander in the field was ignored, we have had no success that can be attributed to our armed forces. At best, we have had stalemates, standoffs, a condition in Korea that is absolutely horrible to consider, where we, for all these years, have had to mount guard on a line of partition that should have never have been allowed to exist. Every time we attempt to answer today's threats of terrorism, we encounter the fact that our special forces are neither given the proper kind of consideration, nor are they in a condition to meet the challenge as it should be met. You cannot leave planning in the hands of civilians seated at a desk, and then hand the final plan to the assault commander for special forces and expect to have success in the encounter. In fact, it seems almost unreasonable to expect men to risk their lives under such conditions.

During this period, since the time I wrote my book *Countdown for Decision,* we seem to have virtually ignored what should be easily recognized as the very pragmatic aims of the Soviet Union. We have spent resources for the enhancement of science in parts of creation that are unlikely even to be reached by human beings and to parts of creation that reveal to us conditions already long past.

Yet we have not put our strength into that part of space which is immediately about this Earth on which we live, that should have been given to that area because of its immediate effect on us here. This relates as well to the matter of defense as it does to offense, and in the present case, it becomes madness not to turn our attention to an effective strategic defense that will make a nuclear holocaust so unsuccessful as to assure that it would not happen. In fact, the objective should be to make nuclear weapons so ineffective, and in fact to cause them to strike back against those who launched them, as to cause this type of weapon to disappear from the world's arsenal. So long as we have and tolerate the existence of nuclear forces, effectively without adequate defense, we are unable to turn our attention to the myriad benefits which are inherent in the whole matter of nuclear science and physics, because in the mind of the public, there is a relation between the peaceful uses of nuclear energy and the uses of warfare and for great destruction. Unfortunately, this relationship is enhanced by the disinformation campaign of our enemies.

The Civilian Bureaucracy

There are many more things that should be considered, but I will leave you with one last thought. We won the greatest war in all of history, World War II, with more men under arms, more spread-out forces, more parts of the globe covered by warfare, than had ever been the case before, and we did it even without the shadow of a Department of Defense. We did it with a civilian bureaucracy at the top of the national effort, that was minimal, to say the least. We had only two secretaries, the Secretary of War and the Secretary of the Navy, and yet we came through that, and it was certainly, in its relative effect and relative cost, no more costly than the peace has become since.

In contrast, in order to *increase*—and that is meant sardonically—the effect of our forces, we have added a civilian bureaucracy, above the secretaries of the several forces, that is tremendous in size, that complicates every decision to be made, and that controls those things which then are publicly charged to the military as their mistakes.

I submit that, once upon a time, our forces knew how to judge what weapons they needed and were allowed to do so and then were prepared, through having their own scientific and technical resources, to see to it that those who produce these weapons, these new systems, did it economically and rightly and without any kickbacks or graft or any of the things that we run into these days and almost consider comically. I ask you to simply consider whether the effects of this tremendous civilian bureaucracy are the right ones for the defense of this country, and whether this extension has truly added to the effectiveness of our forces and their ability to carry out their mission, which is, after all other methods have failed, to win and to protect this country. I ask you to think about this, since I consider it of great importance.

You might look about you where you have military installations, and see that none of the forces are allowed to take care of their own houses. They can't even look after their own homes, and so many of them are allowed to live away from garrisons, even when they first go into the forces. There is not the opportunity to build that kind of comradeship that means a man will not desert a buddy when the chips are down and the choice is win, lose, or die. We might look and see that our forces that are supposed to be deployable quickly and to be able to sustain themselves in the field, are not even allowed to feed themselves at home, and so there has been very little practice for the men and the mess sergeants and the cooks who are going to have to feed them when they go into the field. This phase of our national defense, I am convinced, could yield a sufficient reduction in the resources devoted to it, to provide plenty of power for a really well-organized, top-level approach to a crash program to provide the strategic defense we must have to put an end to the overpowering and continuing of nuclear destruction.

Let us look to our standards and maintain them in this country because, if we do not maintain our standard of freedom and our standard of justice, if they become subordinate to private causes, we have lost that which has inspired our forces in every conflict to do everything they possibly could to stand to the last bitter end and to always come through to preserve our freedom.

Helga Zepp-LaRouche

In response I can only say we will make sure that this message by General Medaris will be distributed by the Schiller Institute all over the world, since I think it is an extremely important message, including his opening remarks. We are totally committed to undoing this injustice which is occurring right now.

ROLF ENGEL

The Soviet View of ABM and Ballistic Missile Defense

The favorite objection made by opponents of the American research program SDI (Strategic Defense Initiative) is that it represents the "militarization of outer space." What, however, are the facts of the matter, facts that no one can argue with?

In August 1957, the Soviet rocket pioneer Sergei Pavlovich Korolyov succeeded in sending a rocket, with a payload of a weight corresponding to atom bombs of that time, over a distance of 9,000 kilometers. Our concept of an intercontinental ballistic missile (ICBM) dates from that time. Since that test, missiles with higher performance, with lighter nuclear warheads, have been developed. These are the heart of the "armaments spiral," which obviously cannot be stopped, in spite of all disarmament negotiations. In the Soviet Union, the ICBM has been the chief pillar of Soviet striving for military supremacy since Stalin's orders in 1946 to develop ground-based long-range missiles.

The Americans have developed a strategic "Triad," consisting of ground-based, submarine-based and bomber-delivered nuclear forces. Over the years, missiles have not only been improved in performance; they have also been made more precise. Each of the two superpowers must expect that the adversary might be able to destroy the available ICBM arsenals of the other side, and thus force capitulation. In order to meet this threat, ICBMs are housed in complicated launch silos, intended to survive a "first strike," and thus provide the ability to paralyze the adversary in a retaliation strike.

If one compares the stockpiles of nuclear weapons on both sides, it seems as if the Soviet Union has gained a clear superiority. According to the Soviet view, however, this impression is deceptive. We cite only "official statements." The old master of Soviet military doctrine, Marshal V.D. Sokolovskii, published his book *Military Strategy* in 1962, which was republished in 1963 and in a third edition in 1969, and has sold about 200,000 copies, serving as the "Bible" for the training of Red Army officers.

The three editions of Marshal Sokolovskii's book were translated and published in the United States in one volume, so that it is possible to follow the changes in Soviet doctrine according to the sections left out, or added, in each of the three editions. Marshal Sokolovskii, of course, treats of the changes in military conceptions determined by the emergence of long-range missiles. He provides a proof for the fact that the impression of Soviet superiority is deceptive in the first edition of 1962:

"Mastery of nuclear energy and the production of ballistic missiles in the Soviet empire will be a blessing for humanity and exploited for discoveries of nature. ... It is evident that the Soviet Union has left the United States far behind in the conquest of outer space. Nevertheless, this lead is exploited by the Soviet Union in a peaceful, scientific way for the benefit of all humanity."

This statement became the classical definition of the purpose of Soviet missiles. Khrushchev, Brezhnev, Andropov, Chernenko have employed the identical formulation, and Gorbachov will repeat it. In the Federal Republic of Germany, many Social Democratic politicians and nearly all of the Greens believe in the same text.

We Are Proud of Our Missile Defense

Hardly had the first ICBMs been launched, when the respective general staffs of both superpowers demanded of their armaments producers, that they develop a defense capable of hitting and destroying the adversary's missiles. By 1959-1960, the technical conception was clear on both sides, and was rather obvious: the destructive effect of nuclear warheads has a large radius. Therefore, it was merely necessary to deliver a defensive missile in the relative vicinity of an ICBM, and then detonate the warheads of the defensive missile. That were sufficient to destroy an adversary's warheads. An atom bomb was to be destroyed by another, one's own, atom bomb.

Work began on both sides on filling out this

logical-sounding conception. In the Soviet Union, especially under the influence of Khrushchev, defensive warheads with large "megatonnage" (explosive power in million tons TNT) were prescribed. The Soviets did not want to rely upon refinements in guidance systems, and thus precision. In the United States, everyone believed in progress in guidance technology, and therefore the solution was sought in defensive missile warheads of lesser megatonnage.

In 1961-1962, the Galosh defensive missile was built in the Soviet Union, and the Nike-Zeus in the United States. Both were conceived as "antiballistic missiles" (ABM), and tested on a large scale in 1962-1963. The Galosh missile carried a 3 megaton warhead, with an effective radius more than five times that of the American system, which had only 0.4 megaton. The Americans had to compensate for the lower effective destructive radius of their system with refinements in guidance technology, which not only cost money, but also time. Thus, the Soviets believed they had solved this difficult problem of defense by means of an ABM sooner than the Americans. Of this, they were very proud, and so, after the grand initial successes in Soviet space flight, this latest grand feat of Soviet missile technology had to be announced to the Soviet people and the whole world. Some quotes:

- Defense Minister Marshal R.Y. Malinovski said on Oct. 23, 1961 at the 22nd Party Congress of the CPSU: "I must report to you that we have successfully solved the problem of destroying missiles in flight."
- Party chief Khrushchev declared to foreign

journalists, in his usual peasant-crude manner, at the "Peace Congress" in Moscow on Sept. 17, 1962: "The Soviet Union is in possession of an ABM missile that can hit a fly in outer-space."

• One year later, on May 12, 1963, the chief of the General Staff of the Strategic Rocket Forces, Marshal S. Biryuzov, said in his speech: "The problem, of destroying enemy missiles in flight, has been successfully solved in our country."

• His deputy, General V. Tolubko, said one day later: "The Soviet Union already has a complete ABM defense system."

• Marshal Sokolovskii declared on Feb. 18, 1965, in his farewell address (he retired): "We have solved the complex problem of hitting and destroying enemy missiles in flight successfully."

• Marshal Rodion Malinovskii, already mentioned, made a speech in Budapest before officers of the Red Army on April 22, 1965, in which he said: "The Soviet Union has recently deployed a new ABM missile of long-range."

• His successor, Defense Minister Marshal A. Grechko, declared in a speech on doctrine Feb. 23, 1970: "Great changes have occurred in the air-defense of our country. We possess weapons systems with which it is reliably possible to hit enemy aircraft or missiles, without regard for their altitude or velocity, and thus at great distance from the target of their attack, which we are to defend."

It is remarkable that, shortly thereafter, the noise about the *Wunder*-Defense-Missile "Galosh" quieted down completely—the missile was still called Galosh in the West, although it had been improved a number of times, and replaced, and its radar system replaced at least twice. What had happened in the meantime?

During the construction of a complete Galosh defensive missile system, it became clear that the kill-ratio fell dramatically when facing attacks, for example, of 100 enemy warheads. If three Galosh missiles, on average, were necessary to destroy one U.S. warhead, the defense was considerably more expensive than the offense. This ABM defense collapsed under massed attack. But in the Soviet Union the leadership of the party and the state—and the entire Nomenklatura assembled in Moscow—took their boasting about their Wunderwaffe seriously, and they categorically demanded that their domicile—Moscow—had to be protected, no matter what. Moscow was therefore surrounded with a ring of 64 Galosh batteries, and the entire defense system officially unveiled in the fall of 1966. That there were many other cities in the Soviet Union to be defended—no one thought about that very much, not even the Soviet delegation, which met with the Americans in 1971-1972 to negotiate a limitation on ABM systems.

In the United States, doubts were voiced very early and loudly about the value of this ABM conception—nuclear missiles against nuclear missiles. Then Secretary of Defense Robert McNamara was a cool, calculating mind, and he was convinced that a Soviet (and, at that time, Chinese) first strike would not be conducted against U.S. highly populated cities, but rather against the silo field of the U.S.A., in which the American ICBMs were housed. Defend one's own ICBMs, that was the motto, for that was the only way to guarantee the new strategy of "He who shoots first, dies second."

In the meantime, the "Nuclear Non-Proliferation Treaty" was signed on Sept. 1, 1968 by the U.S.S.R., U.S.A., and 59 other nations. On this occasion, President Johnson announced, that he wanted to negotiate a limitation on ABM systems with the Soviet Union. But on Aug. 21, 1968, the Soviet Union invaded Czechoslovakia. That put an end to all negotiations. At the end of that year, Richard Nixon was elected President of the United States, and in his first press conference as President declared on Jan. 27, 1969 that he wanted to initiate ABM discussions with the Soviets, in spite of everything. Following tough negotiations, the first disarmament discussions began in Helsinki on Nov. 17, 1969, which led to the SALT I Treaty on May 26, 1972.

Much had happened in the United States. After 10 years of effort, the United States put two men on the Moon with Apollo-11 on Sept. 20, 1969. From February 1965 (when U.S. troops landed in Vietnam) to 1973-1974, the freedom of decision making in the U.S. government was by and large constrained by these events. The only progress made in the area of missile defense was the replacement from 1968-1970 of the Nike-Zeus system with Nike-X, consisting of the Spartan missile with a range of 250 kilometers, and the Sprint missile, both of which were conceived to protect ICBM silo fields.

A discussion about the military value of the ABM conception surely also occurred in the Soviet Union, although there it occurred under the strict secrecy, which the Politburo had imposed on all problems of missile defense from 1967-1968 onwards. Characteristic of this fact is the following section of Sokolovskii's *Military Strategy*, cut from the third edition, i.e., in 1969, but present in the 1962 first edition:

An anti-missile defense system must consist of the following elements: long-range detection of missiles by means of radar (ground-based or air-borne), or other automatically functioning technical aids (for example artificial satellites), in order to secure detection of the missile already during its boost-phase (at the beginning of its ascent); then the coordinates of the missile's trajectory must be determined, following that, the timely alarm and preparations for active defense measures must occur, i.e., fire readiness of anti-missile batteries, disruption of the enemy missile in its trajectory towards its intended target, and finally its destruction on its re-entry trajectory.

The deletion of this paragraph is understandable, for had it been left in the text, the Americans would have asked themselves just what new space weapons the Soviets had in their pockets. That was precisely what the Americans were not supposed to ask.

What Are These 'New Space Weapons'?

First of all, let us establish the fact, that all of the new weapon systems in the American, or in the Soviet, defensive shield, that are being conceived and invited, are not nuclear defense weapons. The purpose is not to destroy nuclear warheads with a nuclear warhead.

If the many friends of peace took their abhorrence of nuclear power plants and bombs seriously, they ought to have staged mass demonstrations of joy in our cities, because, with the SDI of the Reagan administration, nuclear warheads are to be destroyed with non-nuclear defensive weapons. But such things do not interest a "true" friend of peace; all that counts for him is that these plans are American, therefore evil in principle, no matter what the facts of the matter might be.

There is, for example, a considerable number of facts that concern the Soviets' version of the SDI program.

The differences of opinion in the United States concerning the value of ABM systems has been mentioned. The arguments against the concept of using "nuclear weapon against nuclear weapon" were weighty, and ultimately convinced the government. One can assume that the same considerations were entertained in the U.S.S.R. as well, except that there was no public debate in the U.S.S.R. Discussions were going on in the General Staff since 1960, although the military leadership was praising the performance of their Galosh defense in such glowing terms.

Additionally, there is the fact that the ideas about utilizing laser technology originated with Soviet scientists.

This is the reason for the report in the *Military Strategy* by Marshal V.D. Sokolovskii, already mentioned, who wrote: "Possibilities are being investigated, for using a beam of high velocity neutrons, with the effect of detonators of an atomic mass; furthermore, employment of electromagnetic beams, which destroy the nuclear warhead of a missile on its descent trajectory, or deflect the missile from its target. The diverse form of beams, anti-gravity and anti-matter systems, plasma-beams (ball-lightning) etc. are being investigated as means to destroy missiles. The laser (death ray) deserves special attention in this regard, where it is assumed that in the future every missile or every satellite can be destroyed by a strong laser."

This passage was deleted in the succeeding edition (1968), which confirms that, from 1966 onward, the Politburo of the CPSU had exerted strict censorship on all publications on the subject of beam weapons. A review of Soviet scientific literature shows, however, that this censorship was not airtight. Clearly, the Academy of Sciences in Moscow exerted a certain pressure, to enable their scientists to publish their work. One fact came to the aid of the Academy, because immediately after the publication by T.H. Maiman (in *Nature* 1960) on light amplification by means of a ruby laser, the Lebedev Institute for Physics began intensive work on lasers. It was especially the physicists N.G. Basov and A.M. Prokhorov, who reported on the theoretical and experimental work on laser effects in reports of the Institute between 1961 and 1963. This work was viewed so highly by international experts, that the two physicists were given the Nobel Prize for Physics in 1964.

Observation of Soviet activities in the field of directed-energy equipment became more difficult after that. One needed to review the purely expert journals, such as *Soviet Physics* (Uspekhii) and the *Journal for Quantum Electronics* to find out what was happening in the U.S.S.R. International conferences also were very fruitful, particularly in the area of nuclear physics, astrophysics, space flight, and technology of measurement, where it was possible to learn a good many details in discussions with Soviet colleagues. It was quite apparent that Soviet scientists were extremely well-informed about Western scientific literature.

In the following account, only the most important events, which provide insight into the Soviet "SDI" program, are summarized:

- 1963-1965: N.G. Basov and V.N. Grayevskii—both at the Lebedev Institute—publish a series of ground-breaking papers on the gas-dynamic laser, which caused considerable turmoil in the West.

- 1967: in September/October, N.N. Sobolyov provided a comprehensive report in *Soviet Physics* on various types of Carbon Dioxide Lasers, and clearly indicated their military potential

- 1968: the Soviet researcher V.L. Tal'roze from the Lebedev Institute reported on the first chemical laser, and operated it successfully with hydrogen fluoride, and with deuterium fluoride.

- 1969: Basov and Tal'roze, among others, publish their experimental results from electronically pumped chemical lasers.

- 1970: Basov and others operate the world's first excimer laser with krypton-fluoride.

- 1970: the physicist Konyukhov and others report in the Academy of Sciences on further developments in gas dynamic lasers.

- 1971: the physicist V.G. Dudikov from the

Academy in Novosibirsk reports on ion-beam sources for the production of microwaves.

• 1971: professional journals publish the first designs for an electron-beam-pumped x-ray laser.

• 1971: physicists B.I. Stepanov, V. Zhvakin, and Rubanov publish their results on the "four wave mixer" for optical phase-conjugation, which is the fundamental means for correcting the dispersion of a laser beam in the atmosphere. That this discovery is the result of immense research work in the U.S.S.R., becomes really clear only later, when the book by Professor V.E. Zuyev, Director of the Institute for Atmospheric Optics of the Academy in Tomsk, *Laser Beams in the Atmosphere* (1983), is published in English. This book cites 1,231 pieces of literature—of which 850 are Soviet publications—which gives an impression of the extent of research work. At the same time, it is with this background that the tasks of numerous Kosmos satellites becomes clear, which, according to official reports, are all devoted to "studies of the upper atmosphere."

• 1971: a laser power of 300 gigawatts is achieved in fusion experiments for the first time at the Lebedev Institute.

• 1972: the Soviet Defense Minister Marshal A. Grechko leaves himself a loophole, after signing the ABM treaty with the United States. He declares: "This treaty sets no limits on research and testing of means, with which a country wants to protect itself from attacks by nuclear-armed missiles."

• 1973: new designs appear in the professional press for electron-beam-pumped x-ray lasers, and an x-ray laser is described which is pumped by a nuclear explosion. (E. Teller in the United States believed for a long time, that he was the first to propose nuclear-explosion pumping. Contrary to their own better knowledge, the Soviets claim the same thing to this day, and *Literaturnaya Gazeta* denounces Teller for being a "cannibal" and "hater of mankind" for that reason.)

● 1974: the U.S.S.R. exhibits in Duesseldorf and later in Munich solid-state lasers for commercial applications, and sells a number of these instruments.

● 1975: American early-warning satellites are "blinded" over the territory of the Soviet Union. The Pentagon plays the incident down; today, experts are of the opinion that the cause of the blindings was ground-based laser equipment.

● 1979: the first industrial isotope-separation equipment is operated, using a carbon-dioxide laser. No such equipment exists as yet in the West.

● 1980: in Sary Shagan, the largest testing grounds on Lake Balkhash, a pulsed nitrogen-iodine laser is tested.

● 1980: in Troitsk (near Moscow) a gas-dynamic carbon-dioxide laser goes into operation. In addition, work on a large expansion of the institute's facilities is begun.

● 1981: in Kiev, Pedovra, and Komsomolsk, large phased-array radar installations were built, such as only required for missile defense.

● 1982: tests are observed at Sary Shagan, in which falling warheads of ballistic missiles are destroyed (apparently with chemical lasers).

● 1983: in Troitsk, the institute is renamed the "Institute for Laser Technology." The buildings

and halls have room for at least 3,000 people to
work.

• 1984: State and Party boss Chernenko declares in an interview: "One can not limit the number of nuclear weapons today, not to mention
reducing them, without taking effective measures
to prevent the militarization of outer space." A pro-
hibition of the American SDI program is declared
to be the prerequisite for resumption of the disarmament negotiations in Geneva. This has remained the position of Chernenko's successor,
Gorbachov.

Only the most important developments have been
selected and briefly described here. Anyone, who
compares the facts described with present knowledge in the West about directed-energy weapons,
must acknowledge the existence of an extensive
Soviet "SDI" program, even if the Kremlin loudly
proclaims the nonexistence of any such program.
But it is part of a program of political and psycho-
terror, to demand that a research package, that
does not even exist, be forbidden, in order to discuss actually existing ICBM/MRBM weapons systems.

What is the reason for this remarkable Soviet
attitude?

The answer, that the Kremlin has always condemned others for precisely that which they, the
Soviets, possess, or are doing, is in large part correct, but this answer is not fully satisfying. A conversation, which an American physicist, Dr. Roger
Main, conducted in 1983 with the head of the Institute for Laser Technology in Troitsk, more likely
hits the center of the question. Dr. Main had been

in Moscow for seven years as an industrial representative, is himself a physicist, and had experience in industrial applications of lasers. He asked the director of the institute, why the institute was built for 3,000 people to work there, for that is an enormous number of people. The answer was: Yes, by Western standards the number of people is too large, but in the West it is possible to simply buy the many thousands of components and parts needed out of catalogues, and from myriads of firms, and they are delivered quickly, with the correct precision specifications. He explained that he could not do this. In order to fulfill his prescribed program, all of the necessary components and subsystems had to be designed and produced in his own workshops.

This is the heart of the answer to the above-stated question. The entirety of infrastructure of many, often small supplier firms is almost totally lacking in the Soviet Union. To have to make everything oneself means considerable allocation of time, and so also costs. The same observation has also been made with regard to the Soviet space program, which took more time than comparable projects in the West. Moreover, entire branches of industry are lacking in the Soviet Union, such as precision finemechanics, micro-electronics, electro-optics, which is demonstrated in the Soviet shopping lists presented to trade representations of Western countries.

Now, complex weapons systems, such as those in the area of directed-energy weapons, are based predominantly upon such "modern technologies." The planners in the Kremlin surely believed, that, by concentrating their efforts on chemical lasers, initially ground-based (in Sary Shagan), and on particle-beam and laser-beam weapons, they would be able to achieve the goals they set themselves by the mid-1980s. The incontestable success in missile technology, as well as the initial leaps in laser technology, and in the area of particle accelerators, apparently confirmed this evaluation of the planners. But they underestimated, that it is a long and wearisome way, to transfer modern technology from the stage of laboratory testing and demonstrations of principle, into industrial production, in a fashion that meets the rigorous specifications of military applications. After President Reagan announced the SDI program, it dawned on the Soviets, that they would surely lose a race with the industrial power of the United States, possibly supported by Japan and nations of Western Europe. Thus, the Soviet attempt to blackmail the West into pursuing its wish for real disarmament via the price of prohibiting the SDI. So, the new Soviet motto has become:

> Our pride of yesterday,
> We damn today,
> For otherwise the other guy is going to pass
> us by.

The grand hope of the Kremlin leadership is the assemblage of "friends of peace" in the West. These people, after all, do not believe in a Soviet SDI.

PAUL GALLAGHER

The State of the U.S. Strategic Defense Initiative

It is difficult to describe the current status of the American Strategic Defense Initiative (SDI) briefly, with any detail, because it is not a crash program. It has not yet become a crash program, as the Soviet program is, as you have just heard preliminary and conclusive evidence on that in Rolf Engel's paper. Still less is it a crash program of the type that the United States has carried out before, with the Manhattan Project, with the Army Huntsville team, the rocket project, the Apollo Project. It is not even yet a crash program representing the determination to immediately pursue every avenue and every breakthrough that rears its head as a potential means of antimissile defense, which is demonstrated in the history of the Soviet program.

Until we force the U.S. SDI to change its structure and become a program which carries out experiments on a multiple basis constantly, parallel lines of experiments, overlapping lines of experiments, and looks for breakthroughs and immediately develops them—until then, it will be difficult to describe its current status, particularly because the breakthroughs that are occurring tend to be reported somewhat *after* they actually occur.

Before I describe the current organization of the program, I want to give you an idea of how this can be approached from an entirely different basis, by reading a recent speech made in Italy by the director of the SDI, Lt.-Gen. James Abrahamson, in order to convey the potential of how this program could be thought about. This is an excerpt from his speech concerning the SDI:

> One project which will perhaps be possible to realize [because of the SDI] is that of Krafft Ehricke having to do with the illumination of dark regions, including highways [on the Earth] by means of mirrored beams [conveyed from space]. . . . The prospects are very exciting and, apart from considerations of national security, the only limit to the applications of the new technologies (as Krafft Ehricke would say), is our inventive capacity. In fact, the majority of great innovations in the field of production—those innovations which create new markets and form the basis of new industries—is the fruit of technological victories, more than of any specific market demand. And in the future, technological progress will be the cause of even more powerful changes.

The depth objective of the national space

policy, launched on July 4, 1982 [that was the previous Reagan policy announcement], involves the strengthening of the security of the United States, the maintenance of American space superiority, and the exploitation of space for economic and scientific ends. The strategic defense program offers the possibility of satisfying some aspects of this objective, imposing the necessary premises for utilizing, in the best way, the contributions which the private sector is in a position to make.

. . . I like to think of the SDI as an integral part of a new renaissance in space. The science and technologies in the period of the [Italian 15th century] Renaissance were the instruments which man needed to complete his emancipation from the Middle Ages. In the twentieth century, the space program has created the basis for a new renaissance. Our activities in space have created new opportunities for us to expand our knowledge of the universe and improve the quality of human life. . . .

The SDI could become the nucleus of a new renaissance in space, the renaissance of the twentieth century, and would contribute to the generation of very many new technologies. Around such a program, there is being created an alliance with scientific investigators, who form part of both the industrial and academic worlds, and this interdisciplinary quality will remain one of the most notable tendencies we will inherit from the SDI.

So, this gives you an idea of the potential of this program under our impact, and under the impact of the scientific tradition that we have been celebrating throughout this entire conference. That scientific tradition is very directly influencing the SDI, at least at the levels of those who would like to make it more than it is right now. It is currently a program that is making many breakthroughs, and that is limited by budget cuts and by slavish adherence to the ABM Treaty, from pushing those breakthroughs in the direction of prototypes of ABM defense.

Breakthroughs—and Bottlenecks

Here are some of these breakthroughs, briefly categorized:

- **Large, lightweight mirrors,** which have to be both placed in space and also used in multiple modes of deployment. These have to be 10 times larger than previous mirrors, the largest mirror built so far being that for the Space Telescope. And they have to be built very rapidly, they have to be mass-produced. The Mass Production Techniques Program of the SDI is one of the programs in the fiscal 1986 budget that is to be cut by every congressional version of the budget so far.

- **High-performance signal processors.** These are to recognize objects, which are sensed in space and in boost-phase.

- **Powerful beams.** Two of the most important breakthroughs are those that recently occurred in the free electron laser or FEL. Results reported three weeks ago on the free electron laser at Lawrence Livermore, were characterized by the [President's] science adviser, [George] Keyworth, as being results which were not expected—at least by him—until the end of this century. Also, there have been major breakthroughs recently in the development of x-ray lasers, in the focusing.

The x-ray laser, which was previously considered a very powerful, destructive beam, but one which could not be focused like a laser, has now, by all accounts, been focused by the development of lenses to do that. And the very high frequency chemical lasers, known as excimer lasers, have now reached power levels in the range of 5 to 10 megawatts in technology demonstrations, which are the power levels necessary for the SDI.

These kinds of breakthroughs are occurring in the program on a regular basis. They have been made public by the SDI Office, in order to try to fight the attacks on the program, which the Soviets and their assets have mounted.

However, it is a known and publicly acknowledged fact in the press, at least in Boston, that more money is laundered, more illegal drug funds are laundered through the First Bank of Boston alone in a given year, than the amount of money being spent on strategic defense by the United States. And that is only one of the big International Monetary Fund banks in one city in the United States.

It is also the opinion of the Strategic Defense Office, that the budget of $3.7 billion, which President Reagan requested for fiscal 1986, and which the Congress is now cutting, is less than the Soviet antimissile defense budget for 1980, which was the last year, in which any sort of Soviet research in this field was allowed to be publicly disseminated.

That was the point at which the Soviet program went into a completely hidden or secret mode and began to pursue real breakthrough developments, while denying its own existence. At that time, in 1980, the Soviet "SDI Program" had a larger budget than the one President Reagan originally requested for fiscal 1986.

As a result—this is from *Aviation Week* on Secretary [of Defense Caspar] Weinberger's most recent report to the Congress on Soviet military power—the result is, that the Soviet program, in the major laser technologies, is moving into a prototype-demonstrations phase, and leads the U.S. program considerably.

Here is another indication of the same thing: an announcement that the Soviet space station, for which the Soviet heavy-lift booster, a 300,000-lb.-lift booster, is now being prepared. That Soviet space station will have, both in kinetic-energy weapons and directed-energy weapons, the means of its own defense against attack. They recently launched a service satellite for it.

This situation of the Soviet program is rather ironically referred to, in the discussion among the official services, as the fact that the U.S. program is "ahead" of the Soviet program. This gives you a very brief idea of the most cautious way, in which the Soviet program's crash-program aspects, can now be characterized. The U.S. SDI is seen as a hedge against the unilateral breakout of the [May 1972] ABM Treaty, and unilateral development of ABM defense, and therefore total military superiority on the Soviet side. But it has been some time, as you know, since infantry men hid behind hedges in warfare.

The Budget Cuts

The proposed funding of the Strategic Defense Initiative is for $26 billion over five years; $1.7 billion is the level actually voted in the fiscal year we are now in, fiscal 1985.

This is the Defense Department portion of the program. There is an additional allocation to the development of the x-ray laser and other SDI-related technologies by the Department of Energy. The x-ray laser is the major program in this field, which is not funded by the Department of Defense, but by the Department of Energy, and therefore constitutes a small additional expenditure over the Department of Defense budget. And what is being done with that money in the U.S. program right now, is primarily the research on specific technologies for ABM defense. And the development of architecture studies, that is educated guesses, about the way in which these systems will operate on paper.

The request for fiscal 1986 is $3.7 billion, and, according to the President's proposal, the request would then roughly double again for the following year. In fiscal 1988 and fiscal 1989, the amounts of money that would be required in an accelerated program for those years, really have not been thought about very much. And so, the amounts are listed as the same.

The Boost Phase Priority

Not terminal defense, but boost-phase defense, is the priority mission. In order to continue destroying warheads after the boost phase, we require discrimination and tracking of reentry vehicles, sensing, and also determining, which are reentry vehicles and which are decoys. Any and all of the beam weapons and other antimissile defenses have to be survivable themselves; they have to defend themselves. Means of defense, often using beam weapons, have to be devised in order to protect them. And, in addition, interceptor defenses are being developed and tested for terminal defense.

The very widely publicized, successful test, which the U.S. Army carried out last year, was the test using three different kinds of sensors—that is, optical, light telescopes, infrared sensors, and radars on the ground, in combination—in order to pick up the ICBM that was launched from the test range in the Pacific, and in order to track it for the interceptive missile, which was launched and successfully hit it and destroyed it.

Here is a chart prepared by the Fusion Energy Foundation near the beginning of the announcement of the SDI by President Reagan (Figure 1). It indicates the underlying nature of the development of antimissile defense. A crash program must be launched in order to accomplish a revolution in firepower, by accessing a revolution in human technology.

This is a chart of the last 200 years, roughly, with the passage of time over the last two centuries charted on the horizontal axis, and the mastery by the human race, by nations, of coherent-radiation devices, coherent-electromagnetic-radiation devices, of higher and higher frequency, on the vertical axis. The curve shows the higher and higher frequency in controlled radiation devices that have been mastered and used over time. It also shows that these devices have been used, beginning with the generation of electricity, for the first kind of controlled radiation beam used, and then going up through the development of various kinds of lasers since 1960.

You will notice that this continuous line of breakthroughs—which have, in each case, meant a greater power of the nations of human beings to do work of all kinds, and have been associated with military-technology breakthroughs—that this is now just verging on the use of transitions, energetic processes that occur within atoms, rather than within molecules. These transitions actually occur within atoms in order to generate the power of these lasers. At the limit of that technology is the x-ray laser, and then, beyond that, the gamma-ray laser, particle beams, and then the possibility for the human race to use the full range of frequencies of the electromagnetic spectrum, of possibilities, both for revolutionary weapons of antimissile defense, and also for work of all kinds.

Free Electron Lasers

Throughout history, it has been the case that each kind of coherent radiation that was used, could be in only one frequency at one time. However, this can now be changed with the develoment known as the free electron laser. This is the laser, that Dr.

Keyworth was talking about when he said the results
that were expected at the end of this century, were
reported three weeks ago.

This is a laser that can operate at different frequencies, can be tuned from one frequency to another, and therefore introduces an entirely new
level of freedom of the technology involved. It is
a laser, which, as little as two years ago, was considered absolutely not to be a candidate for antimissile defense, because it had never been operated
at more than a few watts of power. It has now been
demonstrated at 200 megawatts of power—that is
200,000 watts of power—two years later, and is now
well in the range of required power for antimissile
defense use. In fact, now, after two years of being
a nonentity and an impossibility, it is considered
the leading candidate for a ground-based antimissile defense laser.

The free electron laser is a chamber, an accelerator chamber, into which an electron beam is
accelerated; that is, it is an electrically powered beam
weapon, if you will (Figure 2). If you have a big
electrical power source, and another one to run
these magnets, you can have one—if we have enough
electricity. It is an electron beam accelerated into
a chamber, and then its path is altered, in what is
called a "wiggling path," by magnetic fields arrayed
in reverse order along the sides of the chamber, so
that the north-south polarity is reversed at very
small intervals. As the electron beam travels down
the chamber, and is bent in that way, it gives off
radiation. By the proper tuning of the device, a
portion of the energy of radiation can be directly
converted into the energy of a laser, which is also

fired into the same chamber immediately afterwards, so that the electron beam is turned into an amplifier for a laser. And this is how the very high levels of power are attained. It also means that you can determine the frequency of the output free electron laser by the frequency of the laser that you use to put into it in the first place. That is, this basic device can be used to amplify lasers of many different frequencies. And by that, it can be used as a multiple-frequency laser.

This figure shows the Livermore version, which is an amplifier. There is a more theoretically advanced technique being developed at Los Alamos—and also in France and in West Germany—in which the laser, the oscillating electron beam, gives off radiation. And, by traveling around an accelerator chamber many times, the radiation that it gives off is eventually focused into a laser beam, without the injection of any input-laser beam. This free electron laser is the direct translation of an electron beam into laser energy, and when it is developed, it will actually constitute a freely tunable laser output—one which can be altered freely without changing the input laser.

This represents a completely new level of freedom of use of the technology over the entire history I showed you before in Figure 1, in which controlled-radiation beams of higher and higher frequency, but always fixed frequency, were being used by man. Thus, in a sense—although this is by no means *the one* technology that should be pursued, that is not what I mean to imply—this is a technology that demonstrates very, very graphically, in its characteristics, that a crash program for antimissile defense will completely change the laws and capabilities of human technology for every form of work that is done by human beings. And it will represent, as General Abrahamson was indicating, the completion of a renaissance in space and in human work, which was begun by the U.S. space program.

Detection, Sensing, and Tracking

For tracking the missile throughout all phases of its flight, many different kinds of sensors are being developed, including optical sensors and infrared sensors, which are deployed in space. Later in the missile's flight, airplanes can come into play as additional sensors and trackers, using both optical and infrared devices, and also using laser-tracking devices. Finally, from the very beginning of the missile's flight, ground-based radars at various points in the United States, or in Europe or elsewhere, are complementing the tracking by other devices in order to get a precise trajectory of the missile. All of these means of tracking have to be used simultaneously, and they have to be linked together by computer systems yet to be developed.

The largest mirror of the type needed, which has yet been fabricated, is a mirror that is going into space next year on the Space Telescope, which is just over 2 meters in diameter. The SDI will be building mirrors of anywhere from 5 up to 10, or even 20 meters in diameter of the same precision.

To indicate a problem in the U.S. industrial capacity to support this effort, when it becomes a crash program: It has been calculated that the United States currently produces 2 square meters per year of mirrors of this required tolerance that can actually be used for the reflection of high-powered laser beams on the targets—2 *square-meters per year!* That is roughly the size of your kitchen table. This is the total national U.S. production capability at the moment for mirrors of the required tolerance for the SDI. The Space Telescope mirror, which is 2.2 meters in diameter, took, I think, six years to build.

There was a program included in the fiscal 1986 budget, which has been cut out so far by the Congress, which was to develop rapid-fabrication techniques for large mirrors of this type, and to proliferate those techniques into industry. All of those industrial-production problems have been at least thought of as technology problems within the SDI, and it will have to be the center of the development of these kinds of techniques.

Just to close with an example of the kind of thing, that I am talking about. During World War II, an industrial complex was built up in just 18 months in Oak Ridge, Tenn. for the Manhattan Project. It became the primary example of that time of a massive industrial complex built by two of the largest industrial firms in the country, and under the direction entirely of scientists and military leaders, who were directing that construction for the urgent purpose of completing the breakthroughs of the Manhattan Project. This is the kind of industrial capability that will have to be developed.

So, with that, I would like to close. As I have indicated, the attacks on this program in Congress, which has already cut the SDI budget by *one-third,* are attacks against a program that is trying to measure up to *what the Soviet levels in this area were five or six years ago!* That is about as accurate an assessment as I can give you of the status of the SDI at the current time.

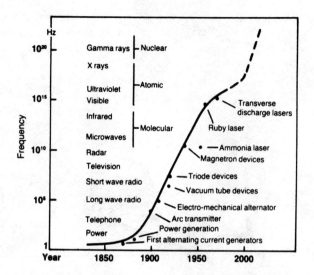

Figure 1 **CHRONOLOGY OF DEVELOPMENT OF SOURCES OF COHERENT ELECTROMAGNETIC RADIATION.** The points on the graph show the date of the first development of devices for generating coherent radiation in the range described in the list to the left of the graph line. The vertical axis shows the frequency in hertz. The development of infrared, visible light, and ultraviolet lasers increased the range of available frequencies of coherent radiation exponentially, a trend that will be accelerated with the development of x-ray and gamma-ray lasers.

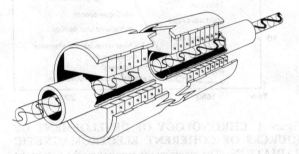

Figure 2 **THE FREE ELECTRON LASER.** A beam of high energy electrons enters the free electron laser together with a pulsed laser field. Magnets arrayed in alternating gradients (north and south) on the device cause the electrons to "wiggle," releasing energy to the beam. At a certain distance along the device, the kinetic energy of the electron is transferred to the input laser beam, thus amplifying it.

DR. NABUKI KAWASHIMA

Japan's Attitude Toward the SDI

I would like to talk a little about the attitude of the Japanese toward the SDI. But first let me introduce myself briefly. I am working on the Space Shuttle, shooting an electron beam, not to shoot down Russian military objects, but for scientific progress. My work is to study the natural phenomena of the interaction of the electron beam with an ionospheric plasma in the atmosphere. When we shoot an electron beam from the Space Shuttle, we can produce an artificial aurora and study this scientifically.

Japan is the first and only country which has experienced the nuclear bomb: Hiroshima and Nagasaki, in 1945, the end of the last, Second World War. And now, the Japanese Constitution stipulates that our country should be neutral and should not have any armaments. So, our people have a very strong reaction to armaments, sometimes a kind of hysterical reaction, especially toward nuclear weapons. So, we cannot develop any nuclear weapons. But our country, of course, is an engineering-developed country now, as you know, so we have the potential capability of constructing nuclear weapons and also the SDI. But when we do that kind of thing, there are very strong reactions not only from our people, but also from our neighboring Asian countries, who recall the bad image of the last Second War.

I think that the SDI is different from other kinds of military weapon developments, because the SDI can lead to the abolition of nuclear weapons, which is our country's goal. Another thing is that it's only a defensive weapon. So my personal opinion is that it's very suitable for Japan to develop the SDI. It is not offensive, and so there should be no opposition from our neighboring countries. I think it should be a new type of peace movement, not by demonstrations with placards, but with the development of new technology, new high technology.

Let me summarize the response of the Japanese people to the SDI. In March 1983, Reagan's announcement was made, but there was no reaction at all. Then, the Fusion Energy Foundation published its book *Beam Defense* in the fall of 1983. I found that book in the airport, and I found it very interesting, so I wanted to translate it into Japanese. Unfortunately, Mr. Kiyoshi Yazawa already had the right of translation. I only joined the translation with some technical advice, and the book was published in Japanese in June 1984. Still, at that time, the reaction of the Japanese was nothing.

The book sold well, and now it has disappeared

from the bookstore shelves. Now, at just the beginning of this year, the reaction has started. Prime Minister Nakasone visited and talked with President Reagan, and at that time, he said that he understood the importance of the SDI. Later on, from April to May, at the time of the Bonn summit, there was a very active debate on beam defense in Japan. So every day in the newspapers and on television, we could see and hear about the SDI.

What will be the final response of the Japanese people is not yet known; it's not settled. Our Liberal Democratic Party, which controls our cabinet, said that it can understand the SDI, but it still hasn't said that it will support or join the SDI. And the other parties, especially the Socialist Party and the Communist Party, it is expected, will oppose it strongly. The party in the middle, and the Socialist Democratic Party, have not decided their attitude, but were a little negative.

I am not a politician, I am a scientist. So I cannot say much about the political and economic response to the SDI. But at this moment, Prime Minister Nakasone has expressed five principles. First, the SDI should not claim one-sided superiority over Russia: There should be always a balance between the United States and Russia. Second, the SDI should be considered within the framework of an overall preventive force against nuclear war. Third, a drastic reduction of offensive missiles should be performed at the same time. Fourth, it should be within the framework of the ABM Treaty. And last, the SDI should be negotiated with Russia when it is actually stationed.

The Weapons Taboo

Now, I must explain something about the background of the environment for scientists in Japan. It may not be easy to understand for you; it's a very peculiar situation. First, for scientists to talk at all about weapons or armaments, or even the *defense* of our country, is taboo. One example is the Japanese Physical Society, which corresponds to the American Institute of Physics. It has about 30,000 members, but no one can be a member who belongs to an organization related to defense. More than that, no papers can be presented which are related to defense-supported research. So, often in international conferences held in Japan, there are many contributions from your country, and in an acknowledgement you write down that the research is supported from some Navy, or Department of Defense, source. Then the Japanese Physical Society requires you to delete this. That kind of thing often happens. Government and university researchers—I am one of them—cannot be supported by any defense grant, directly or officially.

However, I think we have a very strong scientific and engineering capability for the SDI, especially in thermonuclear fusion research. For example, in lasers, we have a very big potential in Osaka University, a glass laser fusion device and also the carbon dioxide laser. The glass laser had been the largest in the world, until recently, just this April or March, when the Nova at Lawrence Livermore started to work. But until that time, the Gekko XII laser fusion machine was the biggest. Also, we have many laboratories doing particle beam fusion re-

search. And although it's not directly related to the
SDI, in fusion research, one of the world's biggest
tokamaks, the JT-60, was completed this spring,
and has started to work. It's competing with the
TFTR at Princeton Plasma Physics Laboratory, and
the JET in Culham, England. We are also constructing a very big accelerator—it's called Tris-
tan—a 30 GeV beam accelerator.

So, in this sense, we have a very strong potential
capability for the SDI. But we are not allowed *explicitly*, in the university or research institute, to say
that we are going to do research related to the SDI.
However, we need to find some "excuse." I will give
you some examples.

Electron beam experiments, electron beam
propagation in the air: This is one of the important
research areas in the SDI, but it has a very nice
scientific purpose. Our Space Shuttle experiment
emits an electron beam, and it goes into the atmosphere. However, when we emit an electron beam
in the *opposite* way, then the electron beam travels
along the magnetic lines of force and it comes back
again, not at the same place, but a little displaced.
And from this experiment, we can see how the
magnetic field is, or how the electric field is, near
Earth. We can study the magnetosphere. This is
one of the major purposes of the Space Shuttle
experiment. But when it is done using the Space
Shuttle, it's very expensive. One Space Shuttle flight
costs $80 million. When we can do this experiment
from the ground, then it's very cheap and we can
do it. The Space Shuttle can fly only once a week
or once in 10 days. But if we can do the experiment

from the ground, we can do stationary observation of space.

Another example is in kinetic energy weapons, or the electromagnetic rail gun. This is not only useful for weapons, but also for studying the impact of a meteorite on the planetary surface, which planetary physicists are very eager to do. Also, it is useful to study high-pressure physics. I am now going to construct a 200-kilojoule machine for a laser gun, which I think will be completed at the end of this fiscal year. In these ways, I think that scientists in Japan can be involved in some sense, not directly, but indirectly, in the SDI.

In conclusion, the Japanese attitude to the SDI is not yet decided; it is changing, it is much in debate. But I think the industries will surely be interested in it: It's high technology.

Maybe by the end of this year, our country's attitude towards the SDI will be decided.

MICHAEL LIEBIG

SDI/TDI: Division of Labor and Complementarity

NATO and the Free World will be confronted with a continuously escalating military threat from the Soviet Union for the remainder of the 1980s. Toward the end of this decade, the total potential of the Soviet Union to conduct and win a global Blitzkrieg against the West will have reached its peak. This goes,

1) for Soviet nuclear offensive forces of all ranges;

2) for the capability of the Soviet Union in operative-tactical, and strategic defense, including civil-defense;

3) for the Soviet classical assault potential—land, sea, and air—particularly against West Europe;

4) for the massive Soviet war-economy mobilization, which cannot be continued indefinitely;

5) for the psycho-political mobilization in the context of the 1,000-year celebration of the Russian Empire.

It is this unprecedented Soviet Russian threat in the next few years ahead, to conduct warfare and/or exert blackmail, which the West as a whole, the U.S.A., West Europe, and Japan, must jointly counter. This will only be possible if a new foundation for the entire Western alliance is laid, in its strategy, and in its personnel and hardware warfighting capability. Defensive beam weapons play the crucial role to that purpose. Only defensive beam weapons can neutralize the spearhead of the Soviet war-plan, its offensive nuclear forces of all ranges. Therefore, defensive beam weapons must be deployed (and can be deployed), in a first generation, however imperfect, by the end of this decade.

Pro-Soviet circles, forces of appeasement, as well as petty-mindedness, shortsightedness, and simple stupidity, have produced conditions in which immense confusion has arisen on the question of beam weapons and the alliance. This awful mess is prevalent in front of and behind the curtains, in public, as well as in political-military and industrial circles.

Since 1982, *EIR* founder Lyndon LaRouche has repeatedly insisted, that research and development, as well as the deployment and strategy of defensive beam weapons, will only be feasible in cooperation between the U.S.A. and her allies. Only in this way, can the inner disintegration of the alliance be prevented, and the conditions for a "crash program" for effective development of defensive beam weapons be established in time.

Unfortunately, up to now an appropriate comprehension of the axiomatic significance of a strategy of cooperation in development and deployment-planning for defensive beam weapons has been lacking in responsible, official circles on both sides of the Atlantic. Under the risk of causing severe damage to the alliance, in the issue of cooperation on beam weapons, these circles have tread the path of pragmatic "trial and error"—one step forward, two steps back—and duplicity.

On condition that there is an affirmation in principle of the goal of defensive beam weapons for defense against nuclear offensive weapons, there are in practice only two paths over which to reach this goal.

The first path consists of that which is already occurring. That means, that in West Europe and Japan, and in other allied countries, research institutions and industrial enterprises are integrated into the American SDI program on a bilateral basis. On account of the scientific and technological capabilities, in comparison to corresponding American potentials, these non-American institutions provide components for the SDI.

The development and production of these SDI components by the allies will lead to a certain, limited flow-back of scientific-technological basic research results in the SDI field, from the United States to the allies. The prerequisite for this approach to allied cooperation is, that the respective governments, whose industries and research institutions work with the SDI, have given the SDI their official or de facto political-strategic endorsement. We can say, with a cautious and sober evaluation of the situation, that this path is being tread already by most of the NATO allies and Israel, albeit with different intensities.

So far, so good, but by far not sufficient!

The path just described, alone and in itself, does not correspond to the conditions and necessities of a crash program for defense against the Soviet nuclear offensive threat, in all ranges, and for the West as a whole. If this path, and only this path is taken, it is more than questionable, whether a crash program appropriate to objective strategic conditions would be possible at all. This approach does not provide us with the necessary efficiency; its working results would come too late; and the material-technological, and personnel base, would remain intolerably thin and inflexible.

The path just described only makes sense, if simultaneously and complementarily a second approach is taken. The point in what follows, therefore, is not "either/or," but rather the one as well as the other. There is no real contradiction between the approach just described and the complementary approach, as long as it is clear, that the goal is to have a first generation of defensive beam weapons, with all of their imperfections, by the end of this decade, for defense against nuclear offensive weapons of all ranges.

This second approach goes beyond component-specific allied participation, and aims at establishing a European Defense Initiative for defense against primarily endoatmospheric nuclear offensive systems in operative-tactical ranges. This second path requires an autonomous focusing of the scientific-technological capabilities of those West European nations, which are capable of the will and resolve, to defend themselves against the Soviet operative-tactical nuclear threat against Western Europe.

The Federal Republic of Germany, Italy and—despite official schizoid ambivalence—France, play the crucial role in this approach. Great Britain will join such an approach, sooner or later, as will the other West European nations.

How the TDI will work

Let me now sketch out the chief characteristics of the European Tactical Defense Initiative (TDI), and its complementary relationship to SDI. In conclusion, Heinz Horeis will briefly describe the main scientific-technological features of the TDI.

1) In the TDI, the scientific-technological, personnel, and material resources of participating West European states, will be brought together and developed, with the aim of generating the means to defend against the nuclear threat against Western Europe, consisting in Soviet operative-tactical missiles, cruise missiles, and aircraft. By the end of this decade, a first generation of these defense systems must be available for deployment.

2) The TDI Organization will work directly and tightly with the SDIO in basic research on beam technologies, as well as the broad field of auxiliary technologies. This close cooperation will run parallel to, and complement, bilateral European participation in SDI on a component basis. This close cooperation is also designed to avoid costly and time-wasting duplication of effort.

3) In the cooperative relationship between SDI and TDI, there is a division of labor differentiation between beam weapon technologies in the endoatmospheric area, which is the primary focus of the TDI, and those technologies for exoatmospheric

deployment, which is the chief concern of the SDI. This natural division of labor, arising from the different threat priorities, will give rise to a fruitful scientific-technological dialogue and mutual exchange of working results and experience.

4) The working principle of the TDIO is that of the "crash program," based on the experience of previous crash programs in this century, such as the Peenemünde, Manhattan, and Apollo programs. The TDIO will be led by a small, General Staff-like group of statesmen, military, scientific, and technical experts from the participating nations.

5) The financing of TDI is to occur outside of regular government budgets, i.e., "off budget," by ensuring that long-term, low-interest credit is extended by state-controlled financial institutions. The TDI will lead to the establishment of new research institutions and high-technology industries, as well as a corresponding increase of high-skilled jobs. Moreover, the spillover benefits of the total productivity increase in society will soon make themselves felt.

6) The TDI cannot replace "classical" armed forces. The maintenance of Europe's classical defensive potential is indispensable, although we can reliably foresee that, by the next decade, the technological characteristics of traditional armed forces on land, at sea, and in the air, will have decisively changed, through the impact of beam-related technologies. For a transition period of a duration that cannot be precisely determined, nuclear weapons too, particularly enhanced radiation weapons, will remain important. This will be so until beam weap-

ons beyond the first-generation will have gained clear dominance over the offensive.

7) By the end of this decade, Western Europe must have a first generation of defensive systems against the operative-tactical nuclear threat against Western Europe. This capacity must be sufficient to render the risk to the Soviet Union of an aggression against Western Europe truly incalculable. This will take the form of a steadily improved layered defense belt of beam weapons, within the atmosphere, linked to the American space-based layered-defense system. The defense layering in Western Europe will consist of fixed and mobile ground and airborne systems, as well as ground/space hybrid systems.

Heinz Horeis will describe the characteristics of such systems.

HEINZ HOREIS

Technical Parameters of SDI/TDI

1) It is completely unfounded to claim that defense against nuclear assault is scientifically-technologically feasible for the United States, but not for Europe. There is *no fundamental difference* between SDI and TDI. The assignment in each case is the same: to acquire adversary missiles as targets, follow them, and destroy them, before they reach their targets. A defensive system capable of doing this will consist of basically the equivalent components, respectively, in the case of SDI and TDI: early warning, target acquisition and following, i.e., sensor technology, power sources, beam technology, battle management.

Therefore, there is a certain "basic requirements" area of R&D, in which technologies equally important for both partners are to be developed, and cooperation between the two partners can decisively accelerate the program as a whole. The fact that officials of the American program have shown considerable interest in technologies developed by West European firms constitutes proof of this.

2) There are, nevertheless, important differences between an American and a West European defensive system, which permit a meaningful and fruitful division of labor to be determined. For the American side, the assignment is strategic defense—defense against missiles and other airborne weapons with long ranges and long flight times, which can be destroyed in space (exoatmospheric defense). Western Europe, on the other hand, is confronted with two forms of nuclear threat:

a) Medium range missiles, such as the SS-20, with ranges of some thousands of kilometers. These can be classified as strategic weapons;

b) The tactical threat—short-range missiles, cruise missiles, aircraft.

3) Defense against "Euro-strategic" missiles ultimately is more or less equivalent to defense against ICBMs, and can rely upon systems developed in the context of SDI. That does not mean, however, that the American screen would "automatically" take over this task. Since we must presume the worst case—a simultaneous assault upon Western Europe and the United States—Europe requires its own system to defend against strategic systems: space-based and/or ground-based beam weapons with long ranges.

4) The chief focus of a European beam-weapon program (and here is the essential difference from the American program) is defense against the tac-

tical threat. In the case of short-range missiles, this means: ranges of a few hundred kilometers, flight times of a few minutes, trajectories within the atmosphere. In orders of magnitude, we will be dealing with some thousands of objects flying in a very few minutes, which puts extreme demands on early warning, target acquisition and following, as well as battle management. Beam defense in Europe will therefore probably be much more "hairy" than for the United States.

5) Three lines of defense against the tactical threat are conceivable, all of which are endoatmospheric:

   a) Airborne high-performance lasers with ranges of some hundreds of kilometers;

   b) Ground-based systems near the borders with medium ranges (10-100 kilometers);

   c) Mobile systems with some tens of kilometers range for point-defense.

6) From the standpoint of criteria and requirements of endo-atmospheric defense, the following chief areas for R&D are determined in the context of TDI:

   a) Propagation of laser and particle beams in the atmosphere, particularly in the lower levels of the atmosphere. (Complex problems arise here, different from those posed by exo-atmospheric and upper-atmospheric beam systems—optical phase conjugation applications, adaptive optics.)

   b) Primary beam generation development. Emphasis should be on development of compact lasers in the multi-megawatt range, for

installation on land, sea, and airborne vehicles; land-based laser and particle-beam systems for ranges of 5 kilometers to 50
kilometers; high-power tunable lasers for all-
weather capability (Free Electron Laser, fre-
quency-shifting devices, etc.).

c) Development of ultra-high-velocity projectile accelerators for endoatmospheric applications, in particular, of magnetic rail-gun
technology, for anti-missile, anti-aircraft, and
anti-tank weapon applications. Aerodynamic
studies on small, high-density projectiles at
ultra-high velocities in dense atmospheres,
target damage studies.

Additional areas to concentrate on are:

d) Development of compact pulsed power
sources for mobile deployment;

e) Development of stabilized platforms,
pointing systems, optics and power supplies
for mobile basing of directed-energy weapons;

f) Acceleration of European development
of satellite and aircraft-based remote sensing
systems for surveillance of Warsaw Pact operations in Eastern Europe, including "instant
warning" of missile launches.

Most of the task assignments mentioned here are
being pursued in the context of SDI, although in
part in other connections, or with lower priority,
so that a fruitful division of labor is inherently possible and desirable. In particular, the tactical field

would be of interest to the American side for point defense and conventional deployment.

7) Europe possesses the scientific-technological potential to implement such a program. The European space program, with the successful development of Ariane, Spacelab, etc., and JET, the largest fusion experiment in the world, demonstrate this fact, and it is also demonstrated by the persistent efforts of the American side to obtain essential contributions to the SDI from certain European industries, especially from firms in Germany.

This potential must be mobilized, so that *all* of the potentials for a crash program can be exploited, as required in view of Soviet military buildup and concrete Soviet preparations for war. The Eureka program, a peaceful R&D program proposed by French President François Mitterrand, cannot be an alternative to the SDI/TDI for Western Europe, unless Western Europe has already surrendered. Even under the condition that Eureka were not merely a political maneuver against European participation in SDI, even if President Mitterrand had meant his proposal for a European civilian "high tech" program seriously, Eureka would be condemned to failure for two reasons. Firstly, there is no reason to assume that the Soviet Union would honor the civilian character of Eureka with a withdrawal of her nuclear missiles and a reduction of her conventional superiority. Secondly, Eureka would never be able to achieve the chief goal expected by its adherents, i.e., to enable Europe to keep step with America technologically. An application- and task-oriented program like the SDI will run far ahead of such a noncommittal program as Eureka, which carries on "research for the sake of research."

Eureka instead of SDI/TDI means the military, or in the best case, the economic defeat of Western Europe.

DR. WILLY BOHN

Europe Must Be an Equal Partner in the SDI

1) Most of the difficulties in the European discussion about SDI are due to the circumstance, that it is hardly known which concrete programs and projects are actually involved in the SDI. Intensive information is required to alleviate this problem, and this will hopefully occur during the present visits of delegations to the United States.

2) Exaggerated political reactions have been caused by the heated discussion about SDI in Europe, which today surely convey a distorted picture of the positions of most European governments with respect to SDI.

3) Aside from the evaluation of the strategic importance of SDI, the prevelant view, especially in Germany, is that SDI is the motor for an enormous technological development, in which one ought to participate.

4) One of the biggest problems the Europeans

have with SDI is derived from the specific threat against Western Europe, which is fundamentally different from that against the United States. The strategic aspect is, therefore, pushed to one side by the tactical aspects: protection against cruise missiles, medium- and short-range missiles. The technological aspects that are important here, in this tactical consideration, are hardly, or only subordinately considered in SDI. An extension of the tactical features could be the European contribution to SDI, in the context of an integrated program with clearly defined goals.

5) If classified areas are excluded as areas in which Europe participates (as indicated by remarks to the press by General Abrahamson), this would represent an obstacle to SDI participation.

6) The American embargo policy with respect to all components in the high-technology area in recent years has created an unfavorable climate for cooperation: Numerous components for lasers can either not be obtained, or can be obtained in third-class quality. This situation has led to individual nations initiating their own development of these components.

7) Recent cooperation agreements in the military area between the U.S.A. and Germany have exacerbated fears of a technological one-way street.

8) Participation in SDI as sub-contractors does not do justice to the wish for cooperation between partners of equal standing.

9) A technological agreement among European nations, with the goal to become a stronger partner for the U.S.A., is very probable.

10) A technological decoupling of Western Eu-

rope from the United States must be prevented under all circumstances. For this reason, the particular geographic position of the Europeans, with the specific threat against Western Europe, should be translated into the technological parameters, and integrated into a modified SDI program.

The Age of Reason in a World of Mutually Assured Survival and Space Colonization: A New Basis for the Western Alliance

Afternoon Panel
June 16, 1985

A New Basis for the Western Alliance

General Wilhelm Kuntner: After President Reagan's speech in 1983, differently than perhaps in Japan, as we heard today, there was quite a questioning period not only in the United States, but in Europe as well. There were two questions raised all the time, and, I would say, they were influenced by propaganda channels from the other side. One question was, "Are these systems makeable, can we make them?" This is something about which normal human beings cannot say very much; this is something for technicians and scientists. But if it is so—and we have become convinced of it—that the Soviets are doing it already, the second question is unnecessary: "Is it rational?" If the other side does it, there is no other choice but to compete with it and do it.

This is one thing.

The second thing is that I would just like to remind you of the schizophrenia of the demonstrations we see in Western Europe, that is, the demonstrations against the stationing of American nuclear weapons in Europe, when the double-track system [of the Euromissile deployment] was going into effect. Now, you have a new development. Those who are against the SDI are now saying, "Why don't we stick to the second strike capability we have already?"

A third thing one should say as well, is that if one sits, not around Washington, but somewhere around Moscow in a one-way information society, people might convince their own population that what we are doing is a threat. They might feel this as a threat from the other side, and here I think far more information will be necessary across the borders as well, to convince people that it is better to have a system which is attacking nuclear objects, and is not killing people.

Forrest Tierson: I want to say one word about European-American cooperation on the SDI first.

There is a feeling on many people's parts in the industrial sphere in Europe, that things aren't moving very quickly, and that's probably true. But one thing the Europeans have to keep in mind is that the original call for cooperation on the part of Secretary Weinberger in Luxembourg specifically asked the Europeans to propose cooperation along very *specific* lines: in other words, to assess their own security needs—which are different, as has been pointed out many times today—and from those different security needs, to research their own strengths in several different industries, several different technologies, and from that, to put forward proposals for cooperative effort with the United States.

Some of that cooperative effort will also be directly coming from U.S. teams, probably fairly quickly now. But we have to remember that this is a two-way street. The Europeans have their own needs, and in a working alliance like this there has to be two-way traffic, so Europeans should be very careful to propose what they think they can do along these lines, and not just listen to us.

Today, you've heard several different reasons for why we should take this idea of a *crash program* forward. Perhaps I could ask, "Govaryu po russki?" Excuse me if my Russian is a little rusty, I don't get a chance to use it much, and I certainly hope I won't have to use it very much in the future either. I was in Rome participating in a conference of the Schiller Institute, about three months ago, and had the opportunity to stand at the Foro Romano. As I stood there at the ruins of Roman civilization, several thoughts came to my mind. One was that civilization is very fragile and we today stand in many ways in a similar situation. I'm not trying to necessarily equate us with the Romans of that time, but civilizations can end very easily, and the barbarians to the east threaten us all again. Those of us in the United States, those of us in Europe, and also our Asian allies, share many basic human considerations having to do with individual rights, freedom, democracy, things that the Soviet empire does not share and has not shared for as far back in time as most of us can remember. That probably is the single most important reason for this crash program we're talking about: That's the defensive aspect. We must defend ourselves against this constant, continual Soviet buildup. We tend not to believe their literature when they tell us over and over again, as Khrushchev said in the early 1960s, they will bury us. They do mean that.

There are two things to make this program work. The first is to make the threat visible to the populations of the Western world. This has to be done over and over again. Attempts are being made through several organizations to do that. However, this has to be done very carefully, and the United States government is not doing that. It is glossing over all the problems that exist now. The second aspect of getting this crash program to work is to provide some vision, also. It's not just a question of responding to Soviet provocation and some Soviet military construction programs.

We have to point out two things along that line. First, what are the very short-term economic advantages to this particular reindustrialization and movement into space in the sense of Moon colonies, colonies on Mars, and that sort of thing? That kind of point could be made very easily when you tie that to the development of power satellites for beaming power to earth, when you talk about metallurgical and crystal-growing experiments that can be done that can provide tremendous new impetus in the area of crystals and micro-electronics. There are crystals that can be taken to produce devices that have tremendous numbers of units per surface area.

One of the ways we can get industry interested is to demonstrate this to them. You get paid back very quickly. Forget about the spinoff for a moment—it's critically important, it's something that will be coming back, but we can show you how to make money in a very short period of time. This is another way of getting cooperation there.

But even more than that, there has to be almost a philosophical sort of Renaissance. It has to do with movement from place to place. I'll speak just for a moment as a biological anthropologist, talking about human evolution in general. Life has existed probably on this planet for three and a half billion years. When that life first arose, the plant was nothing like it is today. It had an atmosphere which was full of ammonia, carbon dioxide, and methane: just wonderful stuff that we'd all love to step outside now and get a nice big smell of. Over a period of several billion years, those organisms made the Earth as it is today. They purified it. We were mentioning doing something like that with Mars earlier today. But these organisms did that to this planet. We share a history with those organisms, we have some of their genetic material in us today.

There was a period when life existed in the sea and moved out of the sea into an new environment, the land environment. And now we're facing another step in our development. We've reached a point where this planet is not big enough for us and we're facing another move out. We can leave this planet which has been our home for, if you want to talk in an evolutionary sense, perhaps three and a half billion years; if you want to talk in terms of us, ourselves, as a species, perhaps a couple of million years, or less. We can leave that now. It's like leaving home. Leaving home requires growing up, it forces growing up and if we intend to survive, that's certainly one of the things that we have to do as well.

Uwe Parpart: I'd like to stress one aspect of the SDI program which has been discussed somewhat in the course of this conference, but perhaps not with sufficient depth. I would at least like to draw your attention to it.

The SDI, of course, first and foremost, is a military program with a specific military task orientation. But, if it is merely conceived as that, it will not succeed. It has to be thought of in its much broader context, both as a scientific program to advance us in critical areas of scientific research, and from the proper standpoint of method. Secondly, and just as importantly, it has to be conceived of as a broad-based reconstruction program for the economies of the United States and West Europe, Japan, generally for the world economy.

It has to be thought of in terms of the dramatic advances in actual productivity and production technologies that could and should be derived from these programs. The type of advances in productivity that we are talking about are not in the order of magnitude of 2% or 3% or 10% per year, which is what under normal and good circumstances one could expect. But, we are talking about productivity advances in the order of magnitude of tens of thousands of percent. If you use high power lasers for welding instead of using the existing systems, you get productivity increases of about 15,000%. If you use high power lasers properly in the machine tool area, you get similar kinds of advances at least several thousands percent productivity increases.

It's that kind of situation that we have to look at, and have to look forward to. The principal so-called spinoff of the NASA program in the 1960s was the computer technology and related areas in semi-conductor physics. These were important, but from the standpoint of our industry, while these have improved management capabilities, and to a certain extent guidance in industrial systems, they have not really produced the in-depth transformation of our economy. The type of systems contemplated in the context of the SDI, have that capability, have that potential.

When you look at the opposition, both by the Soviet leadership and by those in the United States and Western Europe whose whole philosophical direction is "limits to growth," population limitation, etc., the usual argument which one hears—and this interesting little contradiction is something that we should keep in mind—from the same people in the Soviet Union and the United States who argue that these systems are technologically not feasible, that they are destabilizing, and highly problematical.

Well, they can't very well be both. If Soviet Academician Velikov argues that these systems are not possible, then he should promptly inform Mr. Gorbachov and the arms negotiators in Geneva, not to worry and let the United States go ahead with them. If in fact they're not possible, and they are so convinced of that scientifically, why do they make such a fuss? Why do people at MIT and elsewhere in the United States, who have made it their business to show that these systems are impossible, why do they worry about them? From the Soviet standpoint if these systems are not possible, then what would be better than for the United States to try to build them, because we would simply be doing a foolish thing which would help them. No, they are both impossible, *and* destabilizing and dangerous. Therefore, let us draw the conclusion that the military aspect of this is probably not the most important thing that the Soviet Union is worried about. But the potential transformation of the United States' industry and the industries of the allies, to a totally different and new regime of highly productive production technology, is what undercuts and undermines most profoundly long-term Soviet strategic designs.

Lyndon LaRouche: I am very pleased with what I've seen in the conference as a whole. I think that when the papers and other remarks are put together very rapidly as a package, that we will have assembled, in the discussion here, a research paper which will be useful to the participants for their continued work back in what they do in the coming weeks, and also a paper which will be of immense value to people like [SDI director] Lt. Gen. Abrahamson and others in various countries who are concerned with having a good summation of the various aspects of the problem. I think we've done something for which the Pentagon ought to be grateful, both on the account that we've done the intelligence job, not completely, but a necessary step in the intelligence briefing background, and that there are enough orders that have been given here in military form to facilitate that process, that should be done by institutions.

And, secondly, I think we have, or I hope we have, provoked a jealousy to stimulate efforts and emulation of what we have done in other institutions and circles. We haven't said everything we could say on the subject here in this conference, but I think as I review what has happened here, we have stated many things that need to be stated, not merely in isolation, but as a package which will be useful to many governments and institutions in various parts of the world. That pleases me.

Helga LaRouche: Well, I'm obviously pleased if my husband is pleased. Now I want to invite you to ask any questions you may have on your mind to anybody here.

Questions to the Panelists

Question: One of the things that this conference was supposed to be about was space industrialization. We have heard very little about it. Maybe some of the panel members would care to comment on what they feel United States policy toward commercialization of space should be in the private sector.

Uwe Parpart-Henke: I think there are essentially three sorts of obvious steps: 1) the space station is the next step. That's in the planning stage, and there have been agreements signed between the United States, Europe, and Japan on collaboration on that. Beyond that, we have to consider what Krafft Ehricke has proposed in a book which is going to be published soon on the industrialization of the Moon. That's the obvious second step. Beyond that, there has been some talk and discussion about colonization of deeper space, with Mars being the obvious target. I think that provides a general outline of the direction that we should take. All of these efforts should be undertaken in my mind more or less simultaneously. That is to say, one should not proceed sequentially on this, but undertake this as a program in its entirety, from the standpoint of facilitating the earliest accomplishment of all of them.

As far as private versus public efforts are concerned, I think that the division of labor which exists in this regard in the United States and elsewhere is proper. That is to say, the large-scale research and development efforts are properly government and government-related functions, and should remain this way. On the other hand, we have the Japanese space shuttle, Ariane, or potentially in the future the Japanese launchers that are being constructed. You will find that in a few years Japan will have a space shuttle, for which they made the relevant investment decisions of some $9 billion last year. These efforts of commercial launch in various areas, obviously open a tremendous opportunity for industry to consider manufacturing in space, in environments which are zero gravity, or as in the case of the Moon, reduced gravity environments. From the scientific standpoint, all of the potentials have not been investigated. But I think there is a certain directionality to this which is indicated by this sequence. First, the immediate earth environment, then go to the Moon, go to Mars, and see where we can go beyond.

One of the critical problems which I just want to point out in all of this, is something which is dear to the heart of our friend here, Fred Winterberg. When we want to go beyond the relatively close earth environment, we have to consider in particular new propulsion systems. Chemical propulsion systems for deep space exploration are totally inadequate. Here again you see the confluence of areas of research. Probably the largest research development project in the world today on a cumulative basis is the thermonuclear fusion research project. If you combine that with the requirements for new forms of space propulsion, then, in fact, we can even for the next century look beyond the solar system itself for space exploration, and think about the opportunities that would open up.

Question: I'd like to ask Mr. LaRouche about the three laws of thermodynamics, which first of all I really don't understand, either what they are, or what they say. And I'm curious as to what your opinion might be on those three laws, and how they might work.

Lyndon LaRouche: Well, like learning the laws of a dictatorship, once you learn them, you're going to want to forget them. The so-called three laws of thermodynamics are not laws of thermodynamics. That is, they are not physical laws, even though they're taught as if they were to credulous students in secondary schools and universities around the world. These laws were legislated by associations associated with Clausius and Kelvin beginning in the 1850s. They were not based on any experimental evidence. In fact, they were based on direct contradiction of the entire conclusive evidence, crucial evidence up to that time. Then the laws were amplified, so-called, by another German traitor called Helmholtz, and then by a fellow who committed suicide in Duino Castle in 1901, by the name of Boltzmann. Boltzmann discovered that life in the universe is impossible, unless its clock runs backward. So he stopped his clock from running backwards at Duino Castle in 1901.

One must say that the essence of science does not consist merely in that which is observed. The essence of science is the rigorous comprehension of the relationship between that which is observed and the observer. And when somebody compounds a theory, a biologist or physicist stands before a classroom and specifies to you as the laws of the universe three laws which prove that he does not exist, then you have to put a question mark on those laws.

Question: I would like to address my question to that last German gentleman who spoke. I'm thinking of massed Russian artillery of maybe 500-600 guns, 6 inches to 8 inches in diameter, able to throw pretty big atomic warheads, along with tactical aircraft which might be there, so that there might be many thousands of atomic warheads coming in on the battlefield. Does this technology envision acquiring those targets, those short-range targets of just a few miles up in the atmosphere? Does it envision acquiring the target and zapping them before it gets down to the GI on the ground?

Michael Liebig: I firmly believe that, sooner rather than later, exactly that will be possible. But, what we outlined here today concerns—and that's why the question of a crash program is not a matter of preference, taste, or wouldn't it be better—what we are talking about is what can we do up until the end of this decade, until the end of 1988. I think

we brought forward the conception of a crash p
gram, because only a crash program can give us
the kind of preliminary, imperfect, low-kill ratio,
first-generation tactical defense systems which we
need to neutralize the spearhead of the Sokolovskii
second generation war plan, the Ogarkov plan. To
the extent that we are, within the next couple of
years, capable of neutralizing a significant percentage of a nuclear attack (and significant can be
anything from 5 to 25 or 30 percent of the Soviet
nuclear first-strike capability on the operative, tactical effort), to the extent that we are capable of
achieving that, somehow you pull the rug out from
under the Ogarkov Plan. I think that has to be our
priority. Therefore, it is necessary to eliminate these
categories of offensive capabilities. Because once
they are eliminated, I think that ground forces as
such, without an air cover, without a long-range
rocket artillery cover, without the ability of neutralizing the command and control communica-
tions capabilities of NATO, reserve troop
concentrations, all those that are in the first target
category, of the Soviet first strike—I think this will
give us time at least. I think then that in the more
medium-term perspective, these technical problems can be tackled, too.

Uwe Parpart-Henke: It's easier to shoot down an
artillery shell than an ICBM or an IRBM, because
it flies slower.

Lyndon LaRouche: This is important. The point
that Michael Liebig made is also extremely important. The issue here is a very simple one. We know

the Ogarkov Plan, and the reason we know the Ogarkov Plan is not merely because I've been advised, and others have been advised, by various people who have stolen the secrets or know what the secrets are. We've cross-checked with all sources on what the Ogarkov Plan is. We've also looked at the Soviet press, including *Krasnaya Zvezda* and so forth, and followed these things closely. We study the Soviet scientific publications and from these kinds of things, being good detectives, we're smart enough to know that Jimmy Carter was no good before he was elected. So we're able to understand these things.

More importantly, we understand the Ogarkov Plan because we think in terms of war planning. Therefore, as von Schlieffen would do, we put ourselves in the shoes of the Soviet commanders, knowing what the Soviets are. We say "that's a badger." Okay, now a badger is going to prepare war, in terms of the badger's understanding of badger interests. And what the badger is capable of doing is going to be in terms of the badger's ability to understand things.

We know that Ogarkov can be best described as a Dostoevsky successfully graduating with honors from the Prussian general staff school. That's exactly the way to understand him. He's a Cossack of the Pugachov tradition and the Dostoevsky tradition, and he is probably the most brilliant commanding officer ever deployed in the 20th century, after MacArthur. This man can think. He has *Entschlossenheit* of the kind you cannot find around the Pentagon, at least among the flag officers. You have to get down to colonel rank to find that. We know. I put myself with others in the position of being Ogarkov, working from the Sokolovskii doctrine, which I know to be sound, and working from what I know, the capabilities are now in the future. I said, "Okay, if I'm the Soviet commander, Soviet chief of staff, how do I plan to survive and win a war against the United States and its allies?" When I find these informants from government agencies telling me that Ogarkov is doing exactly what I would do, I figure the guy is pretty smart. At least relative to the kinds of policy we see around the United States these days, particularly in our newspapers and politicians. That's exactly what I would do if I were Ogarkov.

Now, therefore, what are we doing? I know the Ogarkov Plan; I look at the operations, and say, "Okay, Ogarkov is rational. As long as the Soviets believe that the Great Russian People and their specific treasures in Siberia will not survive war, they aren't going to start it unless you push them into the corner to do it. As long as you understand the principles of warfare laid down by Machiavelli—and don't push them into an impossible corner, willfully—as long as they believe the Great Russian People will not survive to enjoy the fruits of victory, they won't start the war. They will be typically Russian. They will wait. They wait, until we destroy ourselves. So, therefore, if you want to stop the Soviets, you have to do one very simple thing. You have to convince them that I have the ear of President Reagan. That will stop them. Why? Because, they know that under that condition, if ten Soviet missiles leave their holes in the ground or the sea, the button will be pushed. And every inch, and specifically Great Russia territory, Novosibirsk, and a couple of other things we know the Soviets prize very greatly, will be obliterated, no matter what happens to Europe and the United States. That wins war. It wins by delaying. They won't even start war.

The same thing is true, as Michael Liebig said, on the question of the artillery. That's the case. What also is true is what Uwe emphasized on the question of the vulnerability of the artillery. But the essence of this matter is not what percentage. Will two get through? Of course two will get through. If I have the best defense of the United States imaginable, I would expect 10 percent to get through. Now 10 percent of a full-scale Soviet assault is not something that New York should laugh about. It would be the first major Soviet contribution to slum clearance. War is war. The question of war is not, can you avoid casualties. We'll have casualties like hell, it'll be uglier than any war you've ever seen, no matter how well we fight it. But casualties would be greater because we've become more stupid than in the past.

We don't even have a civil defense any more. We don't even inoculate our children against biological warfare. We're idiots. We should go out and hang the ACLU today, because you won't be able to hang them as war criminals after the war is over, because they stopped you from inoculating your children. They brainwash parents into saying, don't inoculate your children, it's dangerous, which is like condemning your children to death.

The Question of Warfare

Now the casualties, the question of warfare. Warfare is warfare. It's not a game. It's not politics. It's not a political campaign. The only question in warfare—and when you're dealing with the Soviets, this is essential—is, can you guarantee the survival of your nation as a nation. Whether you have 50 percent casualties or 70 percent casualties is not the question in warfare. The question is, can you survive, can your nation survive as an existence? Number two, having survived the war, will you also win it? If you take away from the Soviet Union both surviving and winning, if you make winning questionable and survival almost doubtful, they're not Khomeini, they're not Qaddafi. They will not put the Great Russian People's existence in jeopardy.

And therefore, you have to learn to play the game of military command, you must develop *Entschlossenheit*. You must do war planning. You're going to walk the edge, as the President of the United States is walking the edge tonight on the question of those hostages in that TWA plane. The President of the United States is now confronted with the point that he must take a terrible action, an exemplary action, not a full-scale war, but an exemplary action against international terrorism, against Syria, against Iran. He must. He's going to have to make the decision. Now if he does what I would do, he's walking the tight wire of nuclear war. It won't actually happen if he does it, but he's walking that wire. There are no safe decisions. There are no simple, safe decisions acceptable to politi-

cians without price or risk. You have to learn to
walk the wall of risk. You have to learn to think
like a commander under warfare conditions. Where
your question is not one of avoiding casualties, your
friends are going to be dead, your home town may
be dead. You're going to continue to fight the war,
but you're going to fight it on the basis of surviving
and winning.

We are in a state of war right now. It's not something that might happen. The Soviet Union, according to its own doctrine of mobilization, is
officially in a state of war against the United States.
The war now is simply a point in the Soviet mobilization in which the mobilization itself is complete, in which the command and control is ready,
and then one morning, without any kind of deployment to forewarn you of that, the maximum
Ogarkov Plan may go into effect. If it goes into
effect within 30 minutes to an hour, most of the
United States and Western Europe will have ceased
to exist. Without warning, and under the present
doctrine, the Soviets would win. Not because the
Soviets would have superior technology, but because we have superior stupidity in our command
structure.

On this question, every little detail—can we stop
every weapon, can we deal with every kind of technology—is unimportant. A million casualties, 10
million casualties, are not important when you're
talking about this kind of a war. Our stupidity got
us into this war, we're already in it. We now have
to fight it and hope we can avoid its escalation to
the next phase of actual war fighting. We have to
learn to think in terms as we do in war command,

war planning. We understand the enemy's war plan, we must operate on the basis of understanding how that works. We must buy delay and advantage by every trick possible. If you've bought enough delay and advantage, if you're acting against his expectations of survival, to cause him to delay, then you've bought time. If you've bought time, that's good, if you know what to do with it.

Question: Mr. LaRouche. While you're standing there, let me ask you how I can bring the Third World situation and its resolution back to my congregation, to the grass-roots level, and make them understand what has to be done. I am afraid that the ideas involved would be beyond them.

The problem is, with the time as it is, in your way of thinking about what is now going on in the Third World, that it is the same kind of situation as the war situation that we are facing. How can we make this visible to the grassroots on a level, where we can raise the type of support that we should? I arrived today, and I am impressed. I came only for the day. I left my church to come. I want to take this back, but I am not really sure whether the grassroots, the simplicity, can reach the level and comprehend, and I am asking you: What do you propose . . .

Lyndon LaRouche: The problem is, the law, the unfortunate law of history—you know this, you say you have got a church, you know a good deal about this problem, you know about sheep. This is the problem. People in the United States watch soap opera. They are recreation-minded, with great turnouts for sports events. The society is collapsing. Nobody is interested in what is happening in Washington, even though Washington is the place that is causing the collapse. Thirty percent, forty percent of our people live on levels, in the United States, that are worse than the Third World. Nobody cares, they look the other way. You walk two blocks beyond the FBI headquarters, and you are in hell! In Washington, D.C., in the slums there. A nation is going to hell. You send a child to college, and you get back a six-legged zombie . . . coming in the frontdoor, and the mother says, "Look, I know," to her husband. "I know, John, you are very upset about our child, but we have to learn to adapt to this." That's what they do, don't they? For every condition, they flee into entertainment, they flee into football, or soap opera, or drugs, or trying 37 or 57 varieties of sex that have just been invented. That's what they do. And you say, people are dying in Africa of starvation. "Well, I have to look after my own affairs in my community," they answer. They are totally insular.

This is, unfortunately, not unusual in history, with this kind of animal. This is true of all nations. People in all nations most of the time behave very stupidly. And they do not respond rationally to events. They don't learn anything from experience. Everyone is saying, "I learned from experience." They learn nothing! They were born yesterday, and during the brief time of their existence, they've learned precisely nothing—except how to behave to get the approval of their friends and neighbors, and teachers. *Populations learn*—and come to higher levels—*only under conditions of great convulsions*. And only when they identify institutions of leadership, which they think will lead them to the Promised Land, or something like that, out of the conditions, they suddenly discover, they can no longer tolerate.

Our problem is, has been, a problem at every point of history, it is a law of history to date: Humanity to date has never gotten out of Hell in many parts of the world, and has barely gotten into Purgatory, which it keeps falling out of in most other parts of the world. We have to understand, as Dante Alighieri and others have understood, the character of our fellow man. It is that divine spark in every newborn child, if somebody else doesn't crush it out, before it gets started. But, what we produce is not the divine result. What we produce is something far short of that. We produce something, at best, from the lower ranks of Purgatory. And these populations respond, not on moral judgment, not on reason—your friends and neighbors are not rational, they are irrational. As *you* know, from dealing with the spiritual problems of these people. They will respond to convulsions, when they suddenly erupt and break up in explosions.

Now, you and I, and others, can see the American population coming up to the threshold, about to explode. The farmers, who thought they were making deals, have just found out that they have just gone out existence. People in general don't care about the farmers going out of existence. They say, "I'm not worried about farmers, I get my food from the supermarket." That's our people. That's the American People, this Great Wise Public Opinion! That's what it is. Public opinion will have to be shocked. We are *in* the shock. People are getting

ready to explode. What they are looking for is a
unity of explosion around which they can unify
under clear leadership.

Our function, as I see it, is to provide, at all cost—
and never allowing frustration to cause us to cease
to do it, never to despair—to continuously present
and amplify, the proposed solutions to the crises
of our time, such that, as the explosion of convulsion occurs, as the world falls around people's
heads—it starts gradually, but then more rapidly—
when the building falls on their head, they know
what to do. If the building rots away, they accept
it, and complain about the rent. It is when the building *falls* on their heads, or goes up in smoke, that
they begin to get excited. And the building is about
to go up into flames. As Percy Bysshe Shelley described it, as Friedrich Schiller described it, as many
others have described it, the Great Works of mankind, like the American Revolution, the period from
1776 to 1789 in world history, the great periods
are very short—or have been so far. A great convulsion, a great disaster, instead of making people
behave more like beasts, as it usually does, under
certain conditions, under certain leadership, men
and women are elevated. They become world citizens in their moral outlook. They are concerned
with the universal consequences—three-four generations ahead, and for nations around them—of
what kind of a world they are building. Then we
see man impassioned, and risen to his most enlightened state! In that enlightened state, mankind is
collectively capable of Great Works, of what we call
renaissance.

But except during these renaissance periods, mankind

is a shameful creature morally! And so we stand at the point, as Helga said repeatedly: Our job, our essential job, is to create a renaissance. You cannot create a renaissance merely by infusing a spirit of goodness. It must be concrete. It must address the crises, and the causes of the crises of our time. It must address the injustices of our time and their cause. It must propose concrete, objectively feasible modes of action. First, to solve these crises, or to deal with them, at least. And, second, to cause people in the moment of questioning, when they say: "How could we have become as stupid as we became?" To give them an answer to that question, that is our only hope.

Question: It doesn't seem as if there is going to be in the foreseeable future a General Agreement on Trade and Tariffs with relationship to our deficit not only here, but the existence of the growing tensions between the United States, Europe, and Japan, with reference to the imbalance of trade. The lackadaisical attitude on the part of the Congress to initiate a crash program—what do you see as the portents for the future?

Lyndon LaRouche: You have this fellow going down the Great Lakes in a canoe. He started all the way at Duluth, Minnesota, and he's going along just fine. Then, he's going down the Niagara River and someone says, "you've got to get off that." He says, "look, things have been going on like this for months, there's nothing to worry about."

We're on a roller-coaster ride to Hell right now. That's not saying that 20 years from now, things will get much worse. No. We're not getting through this year intact. Have you noticed entire banking systems collapsing? Do you know that, according to the Federal Reserve, every major bank of the United States, with the exception possibly of Morgan, is bankrupt right now? Do you know that there were several crises where the dollar nearly blew out of existence in the past period, which were covered up by the government? Do you know that about 20 percent of the food supply of the United States went out of existence this year? Do you know the state of our industry, that when that sewer system in your town collapses, the capacity to replace it is non-existent? Do you know what your car is going to look like next year if you buy one? You're lucky to get a spare tire. Do you know what that reported measles outbreak in Maryland was caused by the ACLU, which campaigned to free your children on religious and other grounds from having to be inoculated from diseases, so your child is now able to spread diseases to other children, as well as himself? Do you know that the great Department of Agriculture in its infinite scientific wisdom, led by Secretary Block-head, prescribed as a good diet for American welfare recipients, a so-called Thrifty Budget Diet, which contains a lower protein and caloric content than the Nazis fed their slaves in their concentration camp system during World War II?

Do you know that many people in this country don't even achieve that level of diet?

Do you know that rats are now at their greatest level in American history in our cities?

Do you know that we are about a few percentage

points below the threshold for an epidemic of bubonic plague throughout the United States? How
many of you were inoculated?

Do you know that the diseases in Africa are wip-
ing out people at a rate which almost satisfies the
Malthusians in the State Department—who ought
to be hung at Nuremberg? If we said what we meant
at Nuremberg, we should be hanging our State
Department officials, because they are proposing
to do what we hung people there for.

Do you know that the time of reckoning is now?
There's no bright tomorrow. Do you know if a pros-
titute is run off the streets of Washington, D.C.,
for professional incompetence, she turns up as a
congressional aide(s)?

What I said before: We are now on the edge of
blows which are partial payment for the stupidity
of our people and institutions over the recent period. Those blows are going to strike now. Those
blows are going to cause convulsions.

Forget the Congress. Congress is not a moral
institution. The Congress will go the way prevailing
winds blow it. And you simply have to terrify the
Congress into doing anything good. They are not
reasonable people. They are people who couldn't
make it in prostitution.

Question: I would like to repeat something said to
me in a meeting up in Chicago, as one of the im-
peratives for immediate implementation of the
Strategic Defense Initiative, so that we could end
this MAD situation and get onto important things,
like how big the lakes should be in Africa. What do
you see as the most immediate potential for cou-

pling technology to the Third World's problem, particularly, the SDI technology?

Lyndon LaRouche: We've already implicitly answered this, but I think we can put a key in there, turn the lock, and all the pieces come together.

In the United States, treason is called the State Department, the Agriculture Department, the Treasury Department, the Commerce Department. Patriotism is called the Defense Department. That is exactly the way you have to look at it. The patriots are a little weak, but nonetheless they're there—"generally weak," as we say.

These institutions are not very moral—the people *in* them can be moral, but institutions have a tendency not to be moral at all. They tend to be *amoral,* at best. The State Department is *immoral,* the Defense Department, at best, is *amoral.* What happens, then, if Brazil, Argentina, Mexico, Central America, Africa, Asia, become vital strategic interests of the war planning policy of the United States and the OECD countries? Under those conditions, then we do things to help those countries, *not* out of moral concern; that is, the *institution* doesn't. *We* may have the moral concern, but it happens because the amoral institution, having decided that as a strategic interest, and applying the principles of technology to *solve* that problem of strategic interest, says, "Okay, we're gonna do it." So that we, the United States, will not build a $3.5 billion mass trunk railroad from Dakar to Djibouti for moral reasons. But if they understand that Chad is the *secret* of control of all Africa strategically—

and even geopoliticians can understand *that*—then, the water systems and the rail systems for Africa, and the energy system, become vital military-strategic interests. And on that basis, the states will act.

You must recognize that—as I've emphasized in some other papers—in the history of science, modern economy and science, scientific and economic progress have *not occurred except under conditions of mobilization for warfare.* That is a measure of the stupidity of mankind, that no nation-state to date, in modern times, has actually conducted generalized scientific and technological progress, *except* under conditions of warfare. That warfare is the most rational act of modern man! Only under conditions of military exigency, a sense of convulsion or anticipated convulsion, has our society yet learned to behave intelligently.

There are people who tell you the military are bad. They're crazy. The military institution is the most rational institution of modern society. There are other institutions which are rational, but they don't have power. Only when the military institution achieves power, and it usually gets it under the wrong auspices, *only* under those conditions, do we behave rationally. And when we behave rationally, we will act properly to change the policies toward the developing countries, not for *moral* reasons—that is, *we* may do it, but the institutions won't do it for moral reasons—the institutions will do it *not* for *moral* reasons, but for *rational* reasons pertaining to vital strategic interests as perceived. That's the only way it will happen.

Question: So that it's clear in my own mind, I would like somebody to elaborate on the laser beam. Now, am I given to understand that it will melt down the missiles? And I want to know what happens to the degrees.

Uwe Parpart-Henke: Yes, Mr. LaRouche just suggested, wear hard hats. The important thing about laser beams, or any other kind of beams of that sort, is that they're capable of concentrating an enormous amount of energy over a very short period of time into a very small place. So it is not a question of melting down a whole missile, or something like that. It's good enough to pinch a little hole into it. And other types of lasers, like so-called x-ray lasers and so on, in fact, will not necessarily be destructive of the vehicle as such, but destroy certain parts inside—which will make the thing dysfunctional.

One of the interesting benefits of an SDI is that based on long-range laser capabilities. If you shoot down a Soviet ICBM in its boost phase; that is to say, as long as the missile engine is still burning, before it has reached its ballistic curve—which will inevitably then carry it to the United States—but if you hit it *before* that, it will fall back on the Soviet Union. It will fall back where it *came* from. Whether or not, under those circumstances, the warhead explodes or something, it is then up to the Soviet military planners to make some decisions about it. Boost phase defense is the critical element of the SDI, both for ICBMs and IRBMs and SLBMs and all of that. But once the missile is on its ballistic curve, it will fall down where it's supposed to fall down, presumably in the United States. But falling down, in a disintegrated fashion, or in so many pieces, is much preferable to having it fall down intact. So we will obviously be dealing with a situation where there's a lot of junk flying in our direction, but if it is not in fact any longer capable of exploding, then it is much preferable to the situation that we would otherwise be facing.

Col. Malloy Vaughn, Jr.: I would like to answer that question a little further. This has been my business. I worked on the development of the warheads; I tested them; I've gone through 11 of them already as guinea pigs. And all I can tell you is that on all our work—I headed up a committee for the antiballistic missile defense in January 1957, I went right on through the testing of it—and I can tell you this: that we have actually tested the different warheads that we've designed, and we know the ones the Russians have got; their design is just like ours, they got our design. We did the R&D, and they manufactured them.

Michael Liebig: In large numbers. . . .

Col. Malloy Vaughn, Jr.: But when you go for the implosion weapons, and you have your defense, and these things are always, you know, covered with H-E, with high explosives. When you disturb *any portion* of that warhead, any *portion* of it, you have a warhead failure. There's no way that that can ever go as a nuclear. You have a pile of H-E that falls. If it's 600 pounds—and one of our warheads is 600 pounds of H-E—or if you have 2,000 pounds of H-E on a megaton, what you're going to do is have a *ton* of H-E, like a bomb that fell in World War II in London; or the V-2—let's go back to the V-2—and you've got H-E that fell there. And you'll go out and pick up the pieces of the radioactive material, just exactly like we did in Canada. You remember that episode up there. We went up there in the tundra, we actually found a piece of radio-active material.

We got excited about the particle. The first time I heard of a neutron concept, like the neutron bomb, or neutron shell, we voted for that very quickly, because we knew we had something that could really do the job. But at the same time, we realized that when you've got neutrons up in the air, in the at-mosphere, you've got an ICBM coming through that, they're going to have a dud. We've already said that: that there are going to be duds coming in, if there's a full attack, because all you've got to have is a disturbance of that existing warhead, and you're going to have an H-E fall on you, but you're not going to have a nuclear warhead. The nuclear warhead is so sensitive that, in some of our designs, everything has to go within a measure of a thou-sandth of a second. And if one little portion of it doesn't function correctly, you've got a dud.

Question: Is it possible to doctor the missile so that it will not become a dud?

Col. Malloy Vaughn, Jr.: We've looked into that, and I can't go any further. But all I can tell you is, that when we were into the Zeus, and going after putting in the radiation, trying to get the pilots, we wanted a dead stick deal, we wanted to take care of the Hercules defense. We wanted to get the pilot, hit him with 5 or 10,000 rad, so he would be like a dead stick. He couldn't actually activate the bomb. That's one way we took care of that.

The other way is that we know they may try to have backups, and all the rest. But when you have a particle—that's where the particle *beam* comes into effect so actively—or *anything*, or you get into the neutrons, you get anything to penetrate that skin. Now, if they want to go ahead and spend a lot of money and redesign on hardening, that's another whole ballgame. But normally, anything that disturbs that in flight, you're going to have H-E come in on you, and it won't go nuke. Now, I'm sorry, but I had to give you that, because there's always this uncertainty about what happens if you knock it down. I had the Los Angeles defenses for two years, and we had more nuclear weapons there at night. On a certain night, I had waving in the air on those little monitors up there, higher yields than we ever had over there in the other areas. But the thing of it is, we knew if anything happened, and it fell over there, we'd have an H-E explosion. We would not have a nuclear explosion.

Question: I'm very pleased to come to this Krafft Ehricke Memorial Conference, because I remember Krafft Ehricke when he spoke to us in Gothenburg, Sweden, and I remember the beautiful paintings and the beautiful ideas he painted, talking about the Moon, and developing the Moon. Now, if we're going to use the Moon for development, and if we are going to use it also for defense—now, if the Russians get there first, what would happen? Would we have a shooting war for the Moon, like we have for the United States of America, or the other parts on this planet? This is one of things that has worried me a lot, and maybe one of the panel would like to answer this.

Lyndon LaRouche: I think the answer is simple. It's quite possible, if we live that long.

Question: Lyn, I think you already may have answered my question, unfortunately. However, with this latest terrorist situation, is there any possible way to get around it without Reagan going on national television and saying, "LaRouche was right; we have to have a Grand Design for global development," *after* simultaneously bombing Qaddafi, Assad, and Khomeini?

Lyndon LaRouche: Well, I don't think we have to do that all at once, but the essential thing to grasp is this. The President has got to make a policy decision—I don't know, he may have already made it this afternoon. At one o'clock it started; it's now six hours later; he obviously made some kind of a policy decision, or an indecision, at least, by now. Our State Department—typical diplomatic indecision. The problem he faces is this: In reality, we're in a state of war. Any decision the President were to make, without starting with that assumption, would be, at best, only accidentally correct. You have to start from the fact that the Soviet Union is now in a state of declared war against the United States. Number two: that the deployment of these terrorists against the United States and its allies is *an act of warfare by the Soviet Union.* Everyone is waiting for the Spetsnaz to attack us within 48 hours of the attack, and that would be the signal to get ready, because you can all sleep until the Spetsnaz killed you. Then you would wake up and turn on the alert system, and prepare for war.

It's not going to happen that way. What you have is Farrakhan and Qaddafi, and the whole bunch of these characters are deploying terrorists to assassinate wives. If they can't get the military officer, they'll slit his family's throat, and his children's throat. That'll spread around, that that's going on; then the military will get discouraged. They'll go say "protect us," and say that we can't do anything for you until after you have been killed—which is the usual State Department line, when someone says, "let's go out and get these guys"—preventive measures. Yes, you have to take preventive measures. You can't do it under civil law, but you can do it under an *act of war.* We're in a warfare situation.

Now, you have over a hundred U.S. citizens, approximately—some of whom have been released—involved in an aircraft seizure. At last report, ten or twelve terrorists are aboard that plane in Beirut, which was then, the last I heard, all gassed up, and they fooded up and whatnot, and it was ready to go off in some other direction. [Voice from audience: It landed in Beirut about fifteen minutes ago.] It landed? [Voice: They're on the ground now.] Okay. So now, the situation is this: We've got one American who is dead, and thrown on the tarmac.

But, on this problem, what have you got? You've got a military decision to make. There is *no way,* even by submitting to the demands of the terrorists, of saving the lives of those people on that plane. You have no assured way. Now, what happens? You say, "Okay. Now I'm going to negotiate with the terrorists this one time—to save lives." How many tens of thousands of people have you condemned to death, by trying to save those lives, in that way?

War Against Terrorism

I don't know if our Delta Force people *could* have taken out that plane, if they'd had the chance. I don't know if they're that good yet; there are some doubts about whether they are receiving the training in capabilities to do it. But, if we had the shot— or maybe with the help of the GSG 9, which we might have pulled into this damn thing—it was blown by NBC, the National Bolshevik Corporation—or as they say in Germany, the National Bolshevist Corporation. So, in the present situation, from a military sense, the lives of those American civilians on that U.S. plane, are written off.

The President's problem now, is that he's got to fight *war.* Because there's a shooting war by Soviet surrogates called terrorists, aided by that Soviet agent, Papandreou (and he's completely a Soviet agent—don't anybody kid yourself—only a certain number of personal defects would prevent them from promoting him to a member of the KGB staff. He's not mentally qualified for the KGB, but they own him, nonetheless.) He was complicit in this operation, and he's run by the Soviets; his government is run by the Soviets. It's really a Warsaw Pact

nation, effectively, right now. The President has to
decide to fight war against terrorism by the best
choice of means available to him.

But also, as President, I would have to do one
other thing. Now, legally, we can't declare a state
of war. Suppose I were President; I couldn't declare
a state of war against the Soviet Union on this one.
It's not politically possible. I've got one other resort.
I've got an intermediate step I can use. I can declare
an ugly national emergency. We already fought a
couple of wars we called "national emergencies,"
as undeclared wars. So, it's in that grey area of war
fighting, which operates under special laws of the
President of the United States, unless he's misad-
vised, that he has got to act right now. And whether
he blows up the Kharg Island or not, is secondary.
The primary thing: he must go into this intermediate state of warfare, which is called "national
emergency."

Now, what we have to do, is *kill terrorists*. And
we have to get the command structures and the
logistical structures of terrorism, in the same way
we would in an ordinary war. We are now officially—or should be, if the President is not corrupted or confused and not doing it—we are now
in a status, where the President has to declare, immediately, a standing condition of national emergency on the grounds of *terrorism*. We now must go
into full-scale, escalating, irregular warfare against
terrorists, whom we go after if they are on the
territory of the United States and its allies (with the
consent of our allies), as we would go after armed
spies and saboteurs.

The approximation in international law on this

thing is, if your country is invaded by people, not in military uniforms, but in civilian clothing, who are coming in as agents of a foreign power to kill and perform acts of sabotage in your government, the police forces must *now* operate under the command of irregular military forces. Which means an activation of special forces, or under the FEMA, so that the law enforcement agencies of the United States, including district attorneys, courts, and other institutions are now (in respect to terrorism, should be, if the President has done the right thing) as of this minute (if he's done it) operating on the issue of terrorism under warfare conditions. The military commanders of the irregular forces, whose job is to find, track down, eliminate, capture, interrogate, etc.—could do all the things you do in warfare. And the U.S. military must be deployed as necessary, again with the support of other agencies of ourselves and our allies, to find the logistical roots, the bases, the command structures of international terrorism, including the Soviet surrogates, to punish them, to deter them, to eliminate them, to hamstring them. And that's the answer.

Now, if the President doesn't do that: If you're young enough, maybe your spine will bend enough so you can kiss whatever you consider precious goodbye. Because, if the President continues to back down—as he's been backing down in the recent months to that Donald Regan-Shultz-Kissinger, etc. crowd—and backs down on this one—another one happens, and he still backs down. This is a testing of the nerve of the United States by the Soviet Command. If the United States backs down on this one, the Soviets will read their capability to escalate to something else in some other part of the world.

If the President responds in the proper way on this—and they are deadly afraid of his capability to respond—and if they think he's responding to my advice, they're going to be sure it's right—and they'll back down. It won't happen with any nice, dramatic, political move. It will happen by the logic of events. And the question before the President on the table now is, is he willing to go to emergency powers, and to declare a state of irregular warfare against terrorists, and punitive action against known agents and supporters of terrorism. And that means putting the law enforcement agencies and courts of this country under the supervision of the military in this matter.

Religious Fundamentalism

Question: In light of the situation that has presented itself with these terrorists and so on, what do you tell people who say, "These are all things that are coming true in the Bible?" They always give you this religious line, setting up world events for this Armageddon scenario. And beyond this, what, if any, was the relation between this Charles Hayes Russell, who was the founder, per se, or whatever, of the Jehovah's Witnesses, and Bertrand Russell?

Lyndon LaRouche: No, no. But the Russell family name is not a good one in American history. It was a famous drug-running family, which founded Yale, out of Connecticut, and the Russell family is the family trust, which is now behind the Harriman family. Harriman is the fellow who made Adolf Hitler look like a pro-black. This man is one of the worst racists in the history of the United States— his whole family.

The first thing (I don't want to get into theology, but it's a very simple matter). To any average human being, we have a separation of Church and State. So, somebody comes in with an argument, and says, "I've got a theological argument, based on my interpretation of some religious text, and therefore the United States government's policies should be run by that." You say, "Whoa, buddy! You're in the wrong country." You say, "But there's a God. The founding fathers agree." "Yeah, sure, sure. Well, how do you know this?" I say, "Well, I recommend that you read something, not a religious text in the normal sense, but I recommend you study theology under Nicolaus of Cusa, who was a canon and cardinal in the Catholic Church, and must be considered as somewhat of an authority on this matter."

Now, I recommend—and I say this in the spirit of Krafft Ehricke himself, and obviously he would appreciate it—well, there's some dispute about the authenticity of a number of passages in the Bible (to say nothing about the authenticity of the literacy of some people who read it), and a lot of this stuff is pretty illiterate stuff. They say, now, this passage, if you look at the various translations, from a decent translation in English, which is called the King James Authorized Version, and to some of these modern subhuman languages, which are considered much better—and they begin to introduce what's called symbolic philosophy. And symbolic philosophy was invented by the Whore of Babylon. You might turn to the latter part of the Book of the New Testament, and I say, well, the Whore of Babylon, she's not a good person. So, symbolic philosophy is out; we don't go with this Simon Magus.

But there's another point, much more fundamental. The question of whether God dictated every passage in the Bible to a person, and of whether the translations and the editions of this thing (particularly concerning some of the people who formed these editions), whether they are authentic or not, is debatable. So, therefore, you have to go with Cusa, to something that is *not* debatable.

The question of any book, or any document—someone says, "I have a document here, which proves. . . ."—a document doesn't prove *anything*, except that it's a document. The validity of a document depends, as in any scientific work, on the ability to independently prove the validity of what it says. And a very good document is one, which gives you a key to the means by which you can prove it. Now, where is the the proof of this document? Take a telescope: the laws of the Universe. There is no question about the authenticity of the Creation. It's there. There is no doubt of the authenticity of the method by which you may study this creation, and observe that it has laws, and that you are pretty meaningless unless you are a part of it—a positive part of it. Therefore, isn't it very simple, that, first of all, you are insignificant, except as you contribute lawfully to Creation? Is it not also true, that you have no purpose in your existence, unless you exist to be an instrument of that kind? So, don't worry about cabalistic prophecies, which come from the

Whore of Babylon anyway. Don't try to interpret the Bible from the standpoint of studying astrology, and symbologies. *Learn* from what is undeniably the work of the Creator, and *learn*, from what you *are*, how you become something in that process, and how you make yourself something in that process. And drive yourself by the existential concern, knowing that you are born and you die, within that brief interval in life in which you live, that you'd better use that, because that's all you've got. And you'd better use that to make some permanent contribution, which makes the guy who created this thing happy.

Col. Malloy Vaughn, Jr.: I was out of the room for about 45 minutes, and during that time we did get the news. We saw the plane land in Beirut, Lebanon. We saw the President getting off his helicopter and being met by Shultz and Caspar Weinberger, and a decision is going to be made, I guess. He said that the only thing that's been on our side, is that we've had all this time—three different trips now, back into Beirut—that's a good sign, they feel they're not going to blow it up. I have a question—and you can correct me very quickly, since I was not here during the opening of this—has anyone brought up the very touchy question of classification between our allies on equipment? Has that been answered already?

Lyndon LaRouche: It's been addressed, but no, I don't think it's been answered. [To Parpart-Henke] You addressed it once—on classification and secrecy.

Col. Malloy Vaughn, Jr.: I'm sorry I missed that. The reason I'm very concerned in this, is that, in the past, during the beginning of the early stages of the missiles at Redstone, for five years I was in the position where I had to handle the classification. And I'm the type of individual who wants to keep the classification as low as possible, because that way you can have the most people use it, and you don't have to worry about it. The danger in this country has always been over-classification of documents. It stops everyone—you can't whisper it, you can't show the documents, you can't travel with it. But when you get other countries in, we have a nasty little word called "no foreign." And that "no foreign" goes in there, and it doesn't matter what your classification is as a German scientist or a French scientist. That locks the door. A very simple "no-foreign" goes on it. Now, this cannot be tolerated. The "no-foreign" has got to be thrown out, and never brought into this project, and then, from there, I have—if you want it documented, we can— the best kept secrets in the American forces today. I know three different projects, and the highest classification was "confidential." Forget the classification of "secret," it was "confidential." And that "confidential" warning went right on through, year after year.

And one was a certain type of charge we had in the Army we used, and another one was a very sensitive project that came out of the Skunk Works. By the way, Lockheed did some good work with the Skunk Works with Kelly Johnson, and he came up with the U-2 and the Blackbird. Now, the Blackbird organization was worldwide—in other words, it could bring scientists in from all over the world to work on solving the problem of the Blackbird, the first plane that ever went above Mach 3, and, as you know, is still driving the Russians crazy over there.

Now, the interesting thing about it is, the sensitive thing about it is, "confidential" was the highest rating on that plane, in the performance. And every time TWA or Pan Am was flying round trip to Hawaii, and we took one of those planes off from Ellis Air Force base, and they saw something that looked like a flying saucer go by, and they swore it was a flying saucer—there was no doubt in their mind. We had radar operators all over the United States, whenever they saw it, they said, "A flying saucer—UFO." It went back in the right pattern, it said "UFO flights." And we traced it out, very quickly we knew where they saw it.

Now, when you saw that, we grounded the plane, when this one PanAm flight got to Honolulu, we grounded the plane, took all like you-all into a room like this, we kept you there until we got a special team to fly out there to debrief you. We got your name, rank, and serial number, and at 5 o'clock in the afternoon, we were debriefed that you saw something, and you said, "I didn't really see it." And we treated you like gentlemen, and we said, "This is a confidential matter. It's very sensitive to our government. You've seen it. We want you to have the right to carry this to your grave with you. What you saw, it's confidential, and that's it." And do you know, that year after year, there was not one single leak in the United States, until, unfortunately, President Johnson thought he had to im-press the Russian leaders, and he broke the news to them.

Uwe Parpart-Henke: I spoke yesterday about what used to be called "Operation Paperclip," when the various German scientists were brought over to the United States to help in various projects, which were military projects. But, before "Operation Paperclip" went into effect, the single largest military classification project in the United States' history came about. There were 3 million German documents that were captured and sifted through. Out of these 3 million, about 30 percent, so about a million of these documents, were classified United States military secrets. These papers were basically papers that had been worked on by some of the people who we discussed yesterday, in aerodynamics, in rocket flight, and so on and so forth. After they had been classified military secrets, the German scientists who had been brought over here under Operation Paperclip had a very real difficulty, because they were no longer allowed to look at their own papers, because they were now United States military secrets. And so, also, these German scientists were illegal immigrants, because they had been brought over here by the armed forces, without State Department approval. They had no passports, they had no visas.

So, all of them were loaded into a bus, in Texas on the border, and shipped out of the United States into Mexico. Then they were given visas by the American consul in Mexico, and were shipped back into the United States, now with proper visas attached. And then, finally, somebody had the bright

idea, why don't we quickly make all of them United
States citizens, then at least they will be able to read
their own papers. So, you may think this is all a
joke, but this is the actual history of what occurred
under these circumstances. There's an obvious absurdity in all of this, and there are more details of
this, which I am sure some of the people who were
involved in it would be able to talk about.

I think, in terms of the present situation, the
most serious impediment is not really classification
in the ordinary sense. The most serious impediment is the fact, that the Commerce Department
has a veto power over what kind of collaboration—
military collaboration or anything else—can go on
between the United States and the Allies. That is
to say, what kind of sales can be conducted, what
type of equipment can be transferred, and so on
and so forth. And it is the Commerce Department,
which is, at this point, for example, the most serious
impediment to any significant United States-Japa-
nese collaboration on SDI—the fear being that the
Japanese will learn certain commercial secrets, which
they will turn to their advantage, and United States
industry will be hurt by that. Well, I think we don't
have at this moment all that much to give to the
Japanese that they don't know already. It is more
likely the case, that the problem exists the other
way around.

But the point is, the Commerce Department, in
fact, interferes in these matters in a way which is
totally unacceptable. On the broadest question, I
think, a person who is hardly suspect, I think, of
wanting to give away military secrets to the Soviet
Union, Edward Teller, has argued that, essentially,

there should be no classification in these areas—
that (as was just said) the only people who are hurt
by it, are the United States scientists and engineers,
who are not capable of collaborating in the proper
fashion, because this "need to know" classification
nonsense is, in essence, contrary to the way in which
a scientist has to work. And, therefore, we hurt
ourselves much more than we hurt anybody else
by this.

Lyndon LaRouche: I think what we ought to clas-
sify as top secret, so no teacher can teach it, is logical
positivism, and Cartesian algebra. The biggest
problem we have—Uwe [Parpart-Henke] has inspired me on this one—is that we've been trying to
get secrets into places like Lawrence Livermore National Laboratories, for example. Not out of there,
but *into* there! You know, everyone is saying, "Ugh!
They're stealing secrets from Livermore."

The problem is, there are two things about secrets, about progress. One is scientific method. That
is the most precious of all secrets. Because, without
scientific method, you can't understand in the first
place. The second thing, is production. If you know
it, produce it. The great military advantage, the
great strategic advantage, is to have scientific
method, to use that to assimilate the invention, and,
as you assimilate, begin to produce. Then you're
ahead.

Arnold Ritter: I shall attempt to make some little
additional comment to that problem of being in the
States as an illegal person. It was actually played
this way, that we got a contract back in the home

country, and this contract said that we were willing to accept an assignment on a temporary basis in the United States, and so we were taken—our families could not come along. So we had to agree to take a kind of business assignment in the States for a two-year period, and, if we had proven ourselves useful, our family could come afterwards.

So, first, there was some agreement, apparently, with the immigration authorities—at the time that we were brought in by the military, under guards, the immigration people had to turn around, so we could, behind their backs, then, come inside. And, well, there was a little bit of theater-playing connected with it. And then, initially, we were put in a camp in Ventura, where you cannot go out without escorts. It took initially two years, and then we had to be taken out again, in order to officially emigrate under the so-called "Source Paper," And that meant since I was in Dayton, Ohio, there was a group put on a train, and there was the same thing, they had to cross into Canada and the authorithies again were informed, "don't see them, please." After we were in Canada, then we could start the official papers, sign everything, etc. Then later the guards on the border of Canada, and then, later on, the United States Immigration Authorities, were very happy about seeing some people coming officially to the United States.

I personally intended to make one more comment, which had bothered me quite a bit, as I had mentioned. Dornberger was placed in Wright Patterson Airforce Base, and based on his position as military leader from Peenemünde, was interrogated. And these people who came to interrogate him, had continuously made notes, and notes, and that went on for quite a while. He had terrific insights into what the Russians really are as people, and what they are planning to do. He knew they felt very bad to have physically lost the war. They had been very much damaged, their industries were damaged, so they were really at the bottom.

So he said, they will start right away in building up. And they have taken thousands of Germans, they have taken the whole Peenemünde facilities, dismantled them, and put them somewhere else. But they did not put any of these people into a kind of confined area, but they said, you are now ours, and you work for us. You just watch it, don't do anything wrong. In this respect, they had a really fantastic situation. They had the most modern drawings for the next generation engines.

All these things were recorded, and Dornberger talked about them. At the time when the Sputnik appeared, there was a great shakeup, and, in that movie "Spaceflight," the reaction was shown. But I expected at least the President to say, Dornberger was right, why did I not realize it? No, there was great surprise. I don't know what happened to all these reports, which Dornberger gave. There must be documents about all this.

Starting a Crash Program Now

Question: We have had a lot of discussion about the technical aspects and how to implement these; I just wonder, what the panel, how they feel the implementation would be in regard to control of a project like this, what agencies would be involved, and just how would you structure it, so that you get the flow of ideas moving. If you look at the wartime Manhattan Project, at a scale much smaller than what we are anticipating, how do you keep that moving in a sense?

Lyndon LaRouche: I suggest, we set up two organizations. First of all, we have to have government organizations, but governments today don't work very well; and if we build something else, and government wakes up, it will coopt what we build. But while we are waiting for governments to make the right decision, private agencies have got to make decisions. I mentioned this to a number of you probably before, because some of you here whom we have not talked to, may be able to do something on this. As I laid out, and I admit the paper was very compact—it had to be to be within in the time-frame available, I tried to cover precisely this point, but let me elaborate this one. If you read it afterwards, you will see, there are things buried in there, which had to be included, but which I did not spend much time on.

We are going to do pure science, as I indicated. Pure science means simply taking good practice from scientific research, and instead of having small machine shops and tool shops work as the scientific laboratories, you simply turn over increasing portions of industry as a whole to do the same thing that the scientific machine shop, the tool shop, does with the scientists in ordinary scientific research.

What we have to do is this, internationally—and the Schiller Institute should in a sense be a catalytic kind of sponsoring vehicle to cause this to occur, because it is the only institution now, which is committed to real sovereign cooperation among sovereign states, or peoples of sovereign states, in creating a new foreign policy in the midst of a crisis, which is also a military policy. What we have to do, is, we as a group of scientists and others—I am probably the best economist in the world, so I threw myself in this thing here. (That's a cruel fact. I wish, I were not.) We have scientists, there are other scientists, who were committed, including this precious group from Peenemünde. I don't think we can do it without this catalytic conclusion of the entire force left of the veterans of Peenemünde— we can't do it. That's the treason here on the OSI side.

We have to form an international scientific team, including our friends in Japan with the problems they have, the limitations they have in the way they work. All countries—Ibero-America, countries in Asia. We are trying to save civilization, and the more nations we can get to participate with their own capabilities, the better off we are. We're going to help develop those nations by doing it.

Now we, as a group of scientific teams, have to, by hook or by crook, go to every industrialist in every part of the world who is willing to turn over some portion—forget the budgets, don't talk about budgets, don't talk about money—we are talking about a voluntary effort, a private voluntary effort, where patriotic industrialists, because they are patriots, will contribute under the title of reseach and development, will create nooks and crannies of their industrial capacity, and turn this industrial capacity over to develop certain kinds of materials and instruments, to develop a class of that. Then the scientific group, which will have no inhibitions about a crash program—we don't have to go to the Congress, or some other institution to debate about whether there should be a crash program or not. We privately create the nucleus of a crash program. We coordinate our efforts in the terms of a international ongoing scientific seminar, the same way the Peenemünde group worked, the same way the best of the Manhattan group worked, the same way the best aspects of the aerospace program worked.

And on that basis, we, simply, by sharing ideas about how to make things work, adopt for ourselves the Mars-Moon-Mars mission assignment, as our reference point. And we adopt as our longterm goal, to complete what Krafft Ehricke and others were involved in. We are going to establish the base on the Moon by means of putting a satellite out there which will enable us to get to the Moon and back easily, and then we're going to go on to Mars. And we, as a group of people—privately, as the Oberth group tried to do it—we are going to try to make sure that society gets to Mars 30 years from now, starts the colonization. And take everything in the meantime as a task—we're going to do it.

Now in this context, we know we have to get there alive. And therefore civilization has to survive on Earth. We will take our Mars-Moon-Mars mission assignment project, as a private venture, and we will dedicate that group of people to, as a by-product of our determination, to get the colony on Mars. We would provide, to the governments which want to fight for this thing, to save civilization here on Earth with this technology. We will throw over the fence, into the military, into the governments, whatever we produce that is needed to assist defense. We will campaign and we will organize, on a voluntary basis, citizens, industries, scientists, internationally, to get this thing moving, so that by the time—and the same thing is true of the Oberth group and the Peenemünde case—that by the time government makes the decision to do what we know should be done now, we will say, 'Gentlemen, you know it already is started. We started it.'

So let, in this case, let Lieutenant General Abrahamson have all the support he can get, to do everything that he is permitted to do, and even some things that he is not permitted to do—if it's possible to get by with it. Do the same thing in Japan, the same thing in the Federal Republic of Germany, the same thing in Italy, the same thing in Austria, the same thing in France. As private people, we don't care. We don't have any treaty organizations, we only have a moral commitment to help each other, and to collaborate to survive. So let us, instead of saying, 'Let's just lobby with government,' let us begin to *do* some of it. Let those people who've got tax problems, who need tax exemptions, or want the tax benefits, let them ante in.

Take a German machine tool firm, which is in trouble. It's valuable. Okay, let's get something for it! Let's *save* it! Any firm in the United States which is about to go under, which we need for this effort, any laboratory, let's *save* it, let's help build it up. Let's *fundraise* for it! Let's do *everything* we can within our resources, to get this thing moving. A crash program starts not as a big thing. A big thing is, 'What I want next week, is I want to have five old auto assembly plants, and put in some five axis machines, and build a thousand MX missiles a year.' That's what *I* want to do next week, on a big scale. But what is required: we start with instruments, which are potentially weapons, or auxiliaries to weapons. It's scientific *work* on a large scale. That's all we need is accelerated scientific work, to build some of the things that we already have a general idea of what we can do. We know the scientific principles involved, or we know what kind of experiments we should run to test the scientific principles we don't quite yet have. Build those instruments on a small scale, build up in the plants the materials capacities, the instrument capacities as a resource, so that you can sit in an office someplace in a planning shop, and plan to build an object, and you know that the capabilities exist in these shops to go out and get the parts to put this thing together and make it work.

And you contribute, you have visiting scientists and engineers who visit these places, work with them, collaborate with them, to help them to perfect this technology—to solve the problem of how to make a certain material, whatever problem it is. We have seminars on each of these problems. *Work* on these problems, help them. Somebody will come up with an idea which will help solve the problem. Get the thing started, because you have to go through six months to a year, under the best conditions, on a relatively small but highly diverse scale, even to get the beginnings of the capacity you need. The crash program, then, becomes *called* crash, in the second and third phase, as you go from little, small-scale batch work, into larger scale production efforts.

We have to do it. We don't have to tell *others* to do it. *We* have to organize internationally the patriots—from the military ranks, particularly the retired military ranks, who are more free to be involved in this. From scientists, from industrialists, from engineers, from others. We actually have to go out and mobilize this internationally on a private basis, as an international private commitment to get things moving, get it started. And then, when Lieutenant General Abrahamson calls over one day, and says, 'Hey, look, I got it through.' Or Weinberger. We say, 'okay, fine, we'll come on board. We're ready to work. We've already started.'

HELGA ZEPP-LAROUCHE

In Conclusion

To conclude the Krafft Ehricke conference, I want to make a couple of remarks. It is very clear that we have to go in this pre-war—or actually, war—situation, to a state of enhanced deterrence, against the Soviet Union. But there's also the question, does that lead necessarily to war? Does it mean that when we go for a full-fledged war mobilization, with everything we have, does that mean that this war will actually occur?

Remember that President Reagan made repeatedly, and so did Defense Secretary Weinberger and SDI Director Abrahamson, the offer to the Soviet Union to enter negotiations for joint development of beam defense with the United States. That offer still stands. The Soviets so far have refused even to consider it. What they have done instead, is what you can see at the meeting of the international communist movement, which is occurring today in Paris, recreating the Comintern. And the subject of their discussion is the collapse of the Western capitalist system.

So it's very clear: We have to negotiate with the Soviet Union from an absolute position of strength. And we have to force them, with our launch-on-warning policy, to give up their plan to have world hegemony. The question is, will this work? Is what Lyn once called the protocol for the negotiations of the superpowers, the proposal to *both* implement the SDI *together*, and, at the same time, to give up the idea of dividing the world into spheres of influence, and instead, to develop the Third World together—is this possible? Is this realistic?

I can give you no guarantee. But I can tell you that such a solution—on the one side, total war mobilization, complete development of military strength, and on the other, achieving a peaceful world order—is at least theoretically quite possible, because it would be in accordance with the laws of the universe. And you have to study Nicolaus of Cusa, who actually developed the kind of philosophical system on which such a world order could be based. Among other things, he proposed that every time a human soul, a human being, develops a new technology, this is so precious to mankind as a whole, that no nation has the right to *keep* it, but there should be a pool, so that each nation has access to that science immediately.

This was an idea produced in the fifteenth century, and I think it is what we have to do today. Nicolaus of Cusa said that peace and *Concordantia* can only be possible in a world where every microcosm is developing itself to the fullest. That means that peace is only possible if all nations accept the full development of each nation, and act in the interest of the mutual development of each other, in the same way as peace can only be accomplished if each individual develops to the fullest.

I want to conclude with something which may surprise you: I wrote a poem for Krafft Ehricke. It is in German, and I must apologize, I am German, Krafft was German, and therefore, I will read it in German. But maybe you will get some of the idea, anyhow. I want to dedicate it to Krafft Ehricke.

Concordantia

Wunderbar leuchtet das Bild, und bezaubert die
    schönen Seelen,
Eines Welten Plans gross, würdig der Menschen
    Natur.
Wachsende Harmonien entfalten sich in den
    Gedanken,
Endlos schöpft die Vernunft, stets über Grenzen
    hinaus.
Furchtlos bezwinget der Forscher die weiten des
    wartenden Weltalls,
Freude durchdringet sein Herz, wahr sind die
    Gesetze auch hier!
Hebet den Blick zu den Sternen und denkt wie
    der göttliche Schöpfer,
Denkt in die Himmel euch ein, göttlich, werdet
    auch Ihr.

Oben vom Himmel gesehen scheint die Welt
    noch ärger zu leiden,
Nimmer erträgt es der Mensch, sinnloses
    Morden zu sehen.

Umgekehrt scheinen die Regeln, enger die
    Grenzen zu werden.
Grösser wird nur noch die Not, böser, gemeiner
    der Kampf.
Unsagbar gross ist das Leiden, das Afrikas
    Boden verwundet.
Tränen fliessen nicht mehr, tot sind der Kinder
    zu viel.
Furchtbarer als selbst das Schicksal der
    Beweinten und namlosen Toten,
Ist die Zukunft der Welt, dann, wenn der
    Mensch verlieret an Sein.

Aber, siehe! Da nähern sich edlere Geister und
    bringen
Wohltuhende Gaben zu uns, heilige Kunde
    herbei.
Herrliche Wahrheit bescheren die ehrnen
    Gesetze, und schöner
Glänzt die verheissen herab, nahe, dem offenen
    Herz.
Glück hat die geschändete Menschheit, nur ein
    Gerechter war notwendig,
Schonung erfährt das Geschlecht, Götterfunken
    sind entfacht.
Liebevoll blicken die Himmel herunter, Frieden
    auf Erden
Ist die Berufung·der Welt, Entwicklung Aller,
    ewig Gesetz.

Conference Resolutions

Krafft Ehricke Memorial Resolution

We, the participants in the Krafft Ehricke Memorial Conference in Reston, Virginia, resolve to honor the memory of Krafft Ehricke as a power in rocket science, as a fighter on the frontier of lunar and planetary exploration, and as a model of indomitable optimism concerning the potentialities of mankind. We further resolve to carry forward the heritage of Krafft Ehricke's life's work by promoting, with every energy we possess, the policies that continue the cause for which he stood: the permanent settlement and industrial development of the Moon, Mars, and the other planets, and the guaranteeing of peace in freedom under a regime of Mutually Assured Survival, rendering impotent and obsolete the nuclear arsenals in existence today.

Krafft Ehricke embodied the best tradition of continental European and German science, of the scientific method required today for crash programs in ballistic missile defense and space science. Krafft Ehricke's example lifts our eyes and our minds toward the stars—toward the hope that humanity might cease to squabble in the mud of this small planet, and might mature instead toward a new Age of Reason in space exploration.

In this great enterprise, Krafft Ehricke's invincible will to progress and his lifelong commitment to civilization as the overcoming of brutality and barbarism, must serve us as a precious guide.

Resolution to President Reagan, Heads of State, and Governments of Western Europe, Japan:

We, participants at the Krafft Ehricke Memorial Conference in Reston, Virginia, June 15-16, 1985 from the United States, the Federal Republic of Germany, France, Italy, Austria, Norway, Denmark, Belgium, Japan, Thailand, Mexico, Peru, Argentina, Venezuela, Colombia, and Bolivia, do herewith resolve:

That the Strategic Defense Initiative must be realized in a crash program that draws on the scientific, technological, and industrial capabilities of the entire Western world;

That the Strategic Defense Initiative must be enhanced by the efforts of the allies of the United States to develop and produce the means of securing their own territories against the threat of short and medium range nuclear-armed missiles, and thus such that these autonomous efforts complement the Strategic Defense Initiative.

Our discussion during the Krafft-Ehricke Memorial Conference has enriched our knowledge and understanding of the necessity, that the Western world commit itself to reversing the strategic ero-

sion of our defense capabilities, and we are determined to work together to secure peace and freedom
for our nations.

Resolution in Support of
Dr. Arthur Rudolph

The conference participants endorsed a resolution
calling for urgent White House action to stop the
Justice Department's Office of Special Investigations' witchhunt against German-American scientists, and to restore citizenship to Dr. Rudolph.

Dr. Krafft A. Ehricke

Krafft Ehricke was a graduate of the Technical University of Berlin. In 1942, he joined the V-2 rocket project at Peenemünde, working on nuclear propulsion concepts and V-2 propulsion and deployment.

After the war in 1946, Dr. Ehricke went to the United States to become a jet propulsion engineer for the U.S. Army Research and Development Division, at Fort Bliss, Texas. Later, he became chief of Gasdynamics at the Army Ballistic Missile Center in Huntsville, Alabama.

In 1952, he joined Bell Aircraft, working on glider designs, and in 1954, he became a General Dynamics Design Specialist. He remained with the company until 1962, becoming Director of Advanced Studies, while doing interplanetary-mission and advanced-vehicle design studies for NASA.

In 1965, he became assistant director of the Division of Astrionics at Rockwell.

In 1977, he founded Space Global, a private consulting firm.

In 1982, Krafft Ehricke completed a 10-year study on industrial development and settlement of the Moon.

He published more than 10 books and co-authored more than 50 papers and articles during his lifetime.

His outstanding achievements included:

- The development of V-2 rocket-propulsion technology;
- Participation in the development of the Atlas ICBM vehicle;
- The design of the first successful liquid-hydrogen rocket, the Centaur;

He was chairman of the American Rocket Society Spaceflight Committee, which recommended the formation of a civilian space agency—subsequently called the National Aeronautics and Space Administration (NASA)—to President Eisenhower in 1957.

He was a fellow of the American Astronautical Society (AAS), the British Interplanetary Society, and the AIAA. He has received the AIAA (American Institute for Aeronautics and Astronautics) G. Edward Pendray Award, the American Rocket Society Astronautics Award, the IAF Guenther Loeser Award, the International Aerospace Hall of Fame Award, and the 1984 AIAA Goddard Award.

List of Participants

Dr. Willy Bohn, Federal Republic of Germany
Project leader, Deutsche Forschung und Versuchs Anstalt für Luft und Raumfahrt, Institute for High Energy Physics, Stuttgart

Senator Vincenzo Carollo, Italy Vice chairman, Christian Democratic senatorial group, Senate of the Italian Republic

Dr. Luis Carrasco, Mexico Professor of Astrophysics, University of Mexico, Mexico City

Konrad Dannenberg, United States Consultant, Alabama Space and Rocket Center, Huntsville; former director, rocket motor development, Peenemünde group; former director, Redstone Rocket production, Redstone Arsenal, Huntsville; former deputy program manager, Saturn booster project, which put first men on the Moon; worked with NASA as a space station program manager until 1973; former professor, U.S. Space Institute

Rolf Engel, Federal Republic of Germany Corresponding member, International Astronautical Academy; honorary member, Hermann Oberth

Society; honorary member, German Society for Air and Space Travel; member, Smithsonian Institution; member, American Institute of Astronautics

Paul Gallagher, United States Executive director, Fusion Energy Foundation; coauthor, *Beam Defense: An Alternative to Nuclear Destruction*

Hans Horeis, Federal Republic of Germany Managing editor, German-language edition, *Fusion* magazine; coauthor, *Strahlenwaffen: Militärstrategie im Umbruch*

Dr. Nabuki Kawashima, Japan Professor of physics, Tokyo Institute of Space and Astronomical Science; project participant, electron beam experiment, U.S. Space Shuttle program

General Wilhelm Kuntner (Ret.), Austria Commander, Austrian general staff academy

Lyndon H. LaRouche, Jr., United States Founder and editor-in-chief, *Executive Intelligence Review*; board of directors, Fusion Energy Foundation; member, International Advisory Board, Schiller Institute; chairman emeritus, National Democratic Policy Commmittee

Helga Zepp-LaRouche, Federal Republic of Germany Founder and executive board member, Schiller Institute; founder, Club of Life; editor, *The Hitler Book*

Michael Liebig, Federal Republic of Germany Director, *Executive Intelligence Review*, Western Europe; coauthor, *Strahlenwaffen: Militärstrategie im Umbruch*

General J. Bruce Medaris (Ret.), United States Commander, Army Ballistic Agency, Redstone Arsenal, 1956-1958; commander, Army Ordnance Command, 1958-1960, under which the United States developed the first antiballistic missile defense system, the Nike-Zeus; author, *Countdown to Decision*

Gertrude Nebel, Federal Republic of Germany Widow of Rudolf Nebel, German rocket pioneer

Professor Hermann Oberth, Federal Republic of Germany Founder, scientific theory of rocket propulsion and space exploration

Uwe Parpart-Henke, United States Research director, Fusion Energy Foundation; coauthor, *Beam Defense: An Alternative to Nuclear Destruction*

Arnold Ritter, United States Pioneer in advanced aerodynamics and wind tunnel development; worked with Krafft Ehricke at Convair division of General Dynamics, 1954-1962

Prof. Dr.-Eng. Harry O. Ruppe, Federal Republic of Germany Chair of Space Technology, Technical University of Munich

Jonathan Tennenbaum, United States Director, Fusion Energy Foundation, Western Europe

Forrest Tierson, United States Professor, University of Colorado; member, Space Foundation, Inc.

Dr. Jürgen Todenhöfer, Federal Republic of Germany Member of the German Bundestag, Christian Democratic Union; chairman of the Bundestag

Committee on Disarmament; spokesman of the Christian Democratic caucus of the Foreign Policy Committee of the German Bundestag.

Dr. Friedwardt Winterberg, United States Professor, Desert Research Institute, University of Nevada; author, *The Physical Principles of Thermonuclear Devices*

Kiyoshi Yazawa, Japan Yazawa Science Office, Tokyo; translator into Japanese, *Beam Defense: An Alternative to Nuclear Destruction*

Vice-Admiral Karl Adolf Zenker (Ret.), Federal Republic of Germany Commander-in-chief, Navy of the Federal Republic of Germany, 1962-1968; board member, Schiller Institute